T0220107

Freedom in Mathematics

Pierre Cartier · Jean Dhombres
Gerhard Heinzmann · Cédric Villani

Freedom in Mathematics

 Springer

Pierre Cartier
Institut des Hautes Études Scientifiques
   (IHÉS)
Bures-sur-Yvette
France

Gerhard Heinzmann
Maison des sciences de l'Homme
University of Lorraine
Nancy
France

Jean Dhombres
Centre Koyré, EHESS
Paris
France

Cédric Villani
Institut Henri Poincaré (IHP)
Paris
France

ISBN 978-81-322-3837-9   ISBN 978-81-322-2788-5   (eBook)
DOI 10.1007/978-81-322-2788-5

Translation from the French language edition: *Mathématiques en liberté* by Pierre Cartier, Jean Dhombres, Gerhard Heinzmann and Cédric Villani, © éditions la ville brûle 2012. All Rights Reserved.

Printed on acid-free paper

This Springer imprint is published by Springer Nature
The registered company is Springer (India) Pvt. Ltd.

What is it called,
when the day rises, like today,
when everything is ruined,
everything is pillaged,
and nonetheless the air breathes,
and one has lost everything,
the city burns,
the innocents kill each other,
but the guilty are in agony,
in a corner of the day that is rising? ...
It's called dawn.

—Jean Giraudoux, *Electra*

Foreword

Series 360

The goal that the 360 series, like many degrees and thus proposing a wide-ranging horizon, has set itself may seem too ambitious. Though completeness may not be a realistic aim, the desire to propose a variety of perspectives and to especially accept direct confrontation, via a conversation which, depending on authors and subjects, could turn into a *disputation*, is at the heart of the editorial project. The principle of the series is to gather around a subject four qualified individuals and make them converse with a view of better bringing out the connections in their arguments and questioning.

This original approach is, therefore, deliberately far removed from the dominant modes of scientific "communication"—single-person monologues, usually a scientist, sociological or journalistic investigations, "beautiful books," books with a message—whose qualities and usefulness are assuredly undeniable, by proposing a dialogue around a topic involving science, never in isolation, but together with its relationship to social, economic, ideological, or political issues. In our view, the mere juxtaposition of written texts, a solution often used in publications, cannot render that which arises from the diversity of opinions, and of the various disciplines, intellectual origins, and social practices the protagonists come from. Hence the decision was taken to base the books in the 360 series on a genuine dialogue between authors, in a single time and place remindful of a theater convention. In our opinion, this genuine dialogue, which explains the oral quality of the books, though of course subsequently polished, is what makes all their wittiness and originality and justifies the risk taken by this sometimes acrobatic *modus operandi*.

This series, a recent initiative of the publishing house *La ville brûle,* started with a volume on the Big Bang Theory, the conceptual pillar of astrophysical laboratories, raised by some cosmologists to the rank of a precise theory—even an "ordinary" one, by the desire to challenge—in the last decade or so, as well as the matter of open discussions on its epistemological status that some reduce to a state of speculations (*Le big bang n'est pas une théorie comme les autres,* 2009).

The series followed with climatic changes, by confronting different perspectives from various origins—economy, climatology, and journalism—reflecting the public debates incited by this major issue (*Changement climatiques: les savoirs et les possibles*, 2010). The avatar of quantum theories that are multiple universes enabled us to make physicists, a historian and a philosopher debate (*Multivers, mondes possibles de l'astrophysique, de la philosophie et de l'imaginaire*, 2010). The burning issue of the future of nuclear power after the catastrophe of the Japanese plant in Fukushima Dai-ichi in March 2011 reproduced the configuration of the book on climate, mixing economics, technology, and journalism as evidence that this issue cannot be dealt with by confining it to only one approach (*Nucléaire: quels scéanarios pour le futur?*, 2011).

Freedom in Mathematics

The series follows with this book on mathematics. It confronts the points of view of two mathematicians, Pierre Cartier who was one of the pillars of the famous Bourbaki group and Cédric Villani, one of the most brilliant of his generation, who received the Fields Medal in 2010. Mathematician and historian of science, Jean Dhombres and philosopher of science, Gerhard Heinzmann, both specialists in mathematics, engage in a fruitful dialogue with the two mathematicians, likely to lead the reader to reflect on mathematical activity and its social consequences in mankind's history and in the modern world. Cédric Villani's popular success proves once again that common awareness has assimilated, even if in a very confused way, the major role of mathematics in the construction and efficiency of natural sciences that are at the origin of our technologies. Notwithstanding this, the idea that mathematics cannot be shared remains entrenched, and even branded as a lack of culture laid claim to by some in the media, cultural, and political establishment. Our authors explore three major directions in their dialogue. The very complex relationship between mathematics and reality is a subject of many debates and opposing viewpoints. The freedom that the construction of mathematics has given man is by enabling him to develop natural sciences, as well as that needed by mathematicians to develop their research. The responsibility with which the scientific community and governments should address the issue of the role of mathematics is in research or education policies. Happy reading!

Sylvestre Huet
Journalist for the French daily Libération
specialized in scientific information
Director of the 360 series for the publishing house *La ville brûle*

Contents

About the Authors

Pierre Cartier Academic by training, emeritus CNRS[1] research director, research scientist at the *Institut des Hautes Etudes Scientifiques* (IHES) of Bures-sur-Yvette, former member of the Bourbaki group.

I was born in 1932 in Sedan, a town full of history. I come from two family traditions: engineers on the paternal side, and teachers on the maternal side. My mother was a very cultured, perfectly bilingual, non-conformist woman, combining love for Wagner's music as well as for Debussy and Ravel's music. In fact, a somewhat rough character to whom I owe my knowledge of the German language and my love for music. In other settings, she would have been typical of the Jewish *intelligentsia*.

Despite my provincial background, I followed the path of excellence in my studies: a secondary school in my home town, followed by the *lycée Saint-Louis* in Paris and then the *École normale superieure*, and I obtained an advanced teaching degree and a doctorate in mathematics (1958). After these *Wanderjahre*, I went to Princeton for two years; there I became acquainted with legendary figures such as Robert Oppenheimer and Andre Weil (brother of philosopher Simone Weil). Last but not least, I served in the military for a long time in the marine corps against the backdrop of the Algerian War.

Then I was a professor at the University of Strasbourg from 1961 to 1971. A major period of reconstruction: universities were experiencing tremendous expansion then, mathematics teaching needed to be completely rethought, scientific collaboration with the Germans needed to be developed after the disaster of the Second World War. In addition, the years 1950–1975 were the hey-day of the Bourbaki group, of which I was one of the pillars; there I became friends with famous people like Cartan, Schwartz, Dieudonné, Chevalley, Weil, and with younger ones.

These were also very turbulent years in the history of France: Algeria War, return of De Gaulle, construction of Europe, (worldwide) student revolutions,

---

[1]*Centre national de la recherche scientifique*. Translator's note.

Vietnam war. As a matter of fact, I was also well-trained as an activist: Protestant scouting (like Rocard[2] from and Jospin[3]), progressive catholic circles, Spanish anarchist refugees (after Franco's victory). I remained as a "mathematician without borders" far beyond these years, fighting against colonial wars and dictatorships both of the East and of the West. I was especially involved in Vietnam and Chile, and remain active within the framework of the cooperation agency of the French Mathematical Society (Cimpa).

I have been working in the Paris region since 1971, and I roamed between Parisian academic institutions: University Paris-Sud, *École polytechnique, École normale supérieure.* I am currently emeritus research director at the University Paris-Diderot and a visitor (for indeterminate term) at the *Institut des hautes études scientifiques* (Bures-sur-Yvette). I am not a member of the French Academy of Sciences (a voluntary choice).

My scientific interests are quite diverse (even eclectic), but centered around group theory and mathematical physics. My often cited thesis relates to algebraic geometry, but I also contributed to differential geometry, number theory, combinatorics, numerical analysis, probability and mathematical physics. I wrote a reference book on "Feynman Integrals" (in collaboration with C. DeWitt-Morette, "Functional Integration, Action and symmetries", Cambridge University Press, 2004).

I have supervised more than forty doctoral theses, and I continue to work with some of my students on multiple zeta values and the Galois theory of differential equations.

Jean Dhombres Mathematician and historian of science, emeritus CNRS research director, director of studies at the Alexandre Koyre Centre of the *École des hautes études en sciences sociales* (EHESS), specialist of the mathematics of functional equations and of their applications, of epistemology of mathematics, and of the history of scientific communities and of the spread of scholarly ideas.

I was born in Paris in 1942 in a modest and very structured household, but by force of circumstance with a limited outward orientation, and was lucky to benefit from the "thirty glorious years", and hence never having to ask myself what I should do out of necessity, but only what I wished to do in the world. Having decided to study at the *École polytechnique,* the Algerian War having ended, I chose to take a chance with mathematical research instead of graduating as an engineer, benefiting once again from a remarkable environment in Paris from 1965 onwards.

I worked in functional analysis and through a series of encounters, I joined a small international group of scientists focused on functional equations. In fact it was in Bangkok that my first book was published as a result of a talk I had given at Chulalongkorn University in 1971. Besides many articles, I am happy to have been

---

[2]Briefly Prime minister of France in 1988. Translator's note.
[3]Prime minister of France from 1997 to 2002. Translator's note.

able to work with Janos Aczel, the uncontested leader in this field, having published with him a book on the subject, which has now become a reference, in *Encyclopaedia of mathematics* (University of Cambridge Press, republished in 2008).

Probably as a reaction to my family background as well as to a scientific training centered on France, I wished to see the world, by learning Chinese, by participating in the establishment of experimental classes in Wu Han in China, then by becoming scientific adviser in a French embassy for three years. I became professor at the University of Nantes as early as 1972. Since for a while I was the head of its mathematics department, I was confronted with issues related to teaching and continuing education of high school teachers. Setting up a mathematics education research Institute in Nantes led me to think that getting involved in the history of mathematics would enable me to identify interesting pedagogical methods. This activity has become a major interest and led to the establishment of the Francois Viète Centre in Nantes.

I was fortunate to be elected as a director of studies at the *École des hautes études en sciences sociales* (EHESS) in 1988 and to the Chair of History of Exact Sciences, a title I chose. That same year I accepted the position of CNRS research director to run a laboratory devoted to the history of science in Paris, which now falls within the framework of the Alexandre Koyre Centre, feeling somehow the successor of this philosopher, by line of descent from René Taton and Pierre Costabel.

I am not trying to minimise epistemological influences that may have determined my choice. For the main difference between mathematical and historical practices is that the latter requires almost unending reading, whereas in the former economy and quickness are qualities. I know that I am merging two contradictory options. On the one hand, provocation, in the style of Paul Feyerabend, according to whom all is good as justification in science, and, on the other hand, the patience of the French epistemological school which consists in taking seriously all the justifications given by mathematicians, even the most dogmatic ones, while analyzing them in their cultural context, a context that is also metaphysical, social and political. I think historians still need to better account for the representation of mathematics in human societies in order to grasp how these societies function, by learning how to properly benefit from the computer revolution which gives access to all the texts, from the past to the present-day. It is indeed true that current mathematics are often useful to comprehend past mathematics, and that a celebratory tone, or mere erudite accumulation concerning the latter is of no use for us.

Emeritus since 2007, I am fortunate to be able to continue to lead a seminar at the EHESS, which in the academic year 2013 will be on the question of authority in mathematics.

Finally, let me mention the *Liber amicorum* which has been dedicated to me and which includes a list of publications (*Reminisciences*, vol. 8, Brepols, 2008), and *Une Histoire des savoirs mathématiques et de leurs pratiques culturelles. De l'émancipation à l'âge baroque à la moisson des lumières* (Hermann 2015).

Gerhard Heinzmann Professor of Philosophy at the University of Lorraine, founder of the History of science and of philosophy laboratory – Archives Henri Poincaré (UMR 71117), director of the *Maison des sciences de l'homme* of Lorraine (USR 3261).

After studying mathematics, philosophy and Greek philosophy in Heidelberg, I specialized in the philosophy of mathematics and logic. I am fascinated with crossing boundaries, with the apparent conflict between two rationalities, exemplified by the exact sciences on the one hand, and human and social sciences, on the other. I like to take a new look at ossified aspects of the obvious and the non-obvious, of the finite and the infinite, and I am aware that what was a matter of common sense 30 years ago no longer is today. In short, I am fiercely opposed to the new boundary represented by dogmatism. From this point of view, Henri Poincaré symbolizes as an idol for me: a great mathematician and a great philosopher, he trod a path between rationalism, which bases the explanation of the world on reason and empiricism, which draws knowledge from experimentation. I too also uphold this boundary as it does not separate but brings closer. It is for this very reason that I have accepted to run the *Maison des sciences de l'homme* in Lorraine.

I draw my scientific interests from a dual source, both systematic and historical: in Germany, the challenges faced in the first half of the twentieth century by formalistic, intuitionistic and logistic positions in the philosophy of mathematics have led to the development of the pragmatistic program of the Erlangen school (Paul Lorenzen, Kuno Lorenz). But I owe my historical sensitivity to Jules Vuillemin: I realised then that the pragmatistic and constructive methods used in the philosophy of mathematics to overcome the difficulties faced also stem from or at least resemble the French tradition around Poincaré, as well as intellectual circles more or less closely associated to it (Gonseth, Piaget, Cavaillès, Beth, Bernays, but also Ajdukiewicz, Brouwer and Weyl). This is especially noticeable if Ludwig Wittgenstein and Nelson Goodman are included as go-betweens for these thinkers and dialogical pragmatism. I decided to institutionalize this connection by having Goodman awarded an honorary doctorate in the framework of the Poincaré Archives in 1997.

My research focuses mostly on the use of intuitive and formal elements in mathematics and logic. No one contests that intuition is necessary for invention. However, opinions diverge as to the role intuition may be said to play in the process of understanding and justification in mathematics. Some ascribe a fundamental and irreducible role to it, others just want to exclude it. I am looking for an answer to this dilemma in a *pragmatic semantics*. It is built upon the works of Kuno Lorenz: mathematical realities cannot be conceived independently from their constructions and the latter are not independent from their language description. Instead of considering the construction and the description of objects as two different procedures of mathematical knowledge, they are regarded as two different aspects of the same process of dialogical construction from a common base of actions taking account of their goals and of the given context.

This methodical approach allows me not only to build bridges between analytic philosophy and Poincaré's work, but also—and this is the most important—to include political and social aspects in theoretical considerations.

Cédric Villani Mathematician, director of the Henri Poincaré Institute (IHP) and professor at the University of Lyon, Fields medallist in 2010.

It is impossible to define my origins with any precision, except to say that I am European – with ancestors from North and South Italy, Greece, Corsica, Île-de-France, Alsace and other regions of France. If you add to this that I am the son of *Pieds-Noirs*[4] and a native of Brive-la-Gaillarde, that I grew up in Toulon, that my higher education took place in Paris, and that I consider myself as being from Lyon, you will get a totally confusing but realistic picture of my geographical roots. With these words in way of introduction, a (racial and geographical) fusion pervades my standpoint from the outset.

Cultural fusion has also played a fundamental role in my life. I was born in a mostly literary family, and ended up studying mathematics at university. After entering the *École Normale Supérieure* of Paris, I was glad to find it a place of inter-disciplinary scholarship and to discover the intense cultural life of Paris, going to concerts, theaters and films whenever I wanted. When writing my thesis, I summarized the influences of Pierre-Louis Lions, Yann Brenier, Eric Carlen, Michel Ledoux; I later found my chosen field in the fusion between analysis, probability and geometry

The ability to rebound to grasp opportunities, candidly trusting, is another concept that means a lot to me. My choice to work in analysis when at heart, I was initially an algebraist; my choice of a thesis on the theory of Boltzmann's equation, when I hardly had the temperament of a physicist; my first collaborations with Italian and German scientists, exploiting ideas put forward during travels or con-ferences; my shift towards optimal transport after a series of coincidences and a beautiful talk given by Craig Evans in Paris; then my shift towards geometry at an impromptu meeting with John Lott in Berkeley; and finally my research on Landau damping... to a large extent all this has been the result of chance encounters and travels. Most of my work have been done in collaboration with French, Italian, German, Austrian, Spanish, Swedish, American scientists... When I was awarded the Fields Medal in 2010, it was obviously in the name of community that sustained me and allowed me to develop, and of my main collaborators – Laurent Desvillettes, Clément Mouhot, Felix Otto, Joh Lott...

Transmission is also one of my fundamental values, perhaps inherited from my family. As I am convinced that teaching is inseparable from research, I have always worked as a university lecturer, and have never applied for a full-time research position. I invested considerable effort in writing reference books and course notes – on optimal transport, measure theory, and kinetic equations. I spent an equal amount of effort in communicating with society as a whole: forums, radio and

---

[4]People of European, especially French ancestry, who were settled in North Africa, in particular Algeria, until these countries gained independence between 1956 and 1962. Translator's note.

television programs, public talks, etc. In my first book for a general public, *Living Theorem*, published in 2012 (in a literary collection!), I describe the daily life of a mathematician in the style of a logbook mixing various forms and influences.

Another key-word, structuring: I am amongst those who regard administrative tasks and governance as noble and useful, assisting intellects and projects. I was formerly the president of the students' association I belonged to, I am now director of the Henri Poincaré Institute, which each year welcomes hundreds of invited scientists from all fields of mathematics and theoretical physics, and from all continents. Outside of mathematics, I am also the president of an association that combines art and technology aimed at the integration of disabled youth, the administrator of a philanthropic initiative in the field of education, the president of a scientific council of a research institute located in Senegal, and the vice-president of an association promoting a federal Europe in the public debate. I like to think that this European commitment is in line with the great mathematician Henri Cartan, both a precursor and a model.

After a mixed background, the ability to rebound, collaborations, transmission and structuring, I will add a last slogan: openness through specialization. As a mathematician, I am fortunate to be in a profession enabling me to travel throughout the world (35 countries to date), be entrusted with numerous responsibilities, come into contact with all sections of society – for science concerns everyone, and everywhere in the world, scientists are at the heart of society. If mathematics as a subject can be a wonderful way to explore the physical world, as a profession it can also be a wonderful way to explore the human world.

On the Origins of Mathematics

Euclid's Elements: The Hard Core

Sylvestre Huet: Why and how did man start doing mathematics, in particular based on the most elementary mathematical objects such as the point, the line, or the surface? Issues about the relationship between mathematical objects and reality arise from the onset: why and how does one do mathematics, and what is the relationship between mathematical objects and real objects or natural sciences? The informed general public is usually aware that the most basic mathematical notions, including numbers, were difficult constructions: think about the time taken to invent zero or positional number systems, concepts that we now learn as early as primary level... Yet, the invention of the concepts required bright minds, the brightest of the time. The relationship between mathematics and reality, which may seem obvious, is not so at all. Jean Dhombres could expand upon that, in order to begin the discussion by its historical aspect.

Jean Dhombres: Assigning a beginning to mathematics depends on what we include in the word mathematics, which is not very old, which attests to the differences in emphasis. Because it is not often underlined, from the outset I start by pointing out that an expression like "a course of mathematics" does not appear before the 17th century as a title for a book. I do not, however, think that our discussion would profit much from committing ourselves to a definition ab initio. Let us instead start by one of the main topics that we are going to address: there can only be mathematics when there is both a form of abstraction and a claim of some connection to reality which should be described as the result of an action, or an operation. This applies to numbers: we are already doing mathematics when we think that "three gazelles" and "three days" have "something" in common, and that "this something", though distinct from a pair or from a single, can be deduced by operations and radically differs from the idea of "many." It is, therefore, obvious that at this stage mathematics is inseparable from the process of language formation, and that abstraction is then supported by writing. But history shows that there are different written number systems, hence cultural factors

© Springer India 2016
P. Cartier et al., *Freedom in Mathematics*,
DOI 10.1007/978-81-322-2788-5_1

do not bring the same operations into play. Positional number systems bring into play addition, Roman numeration brings it in differently by mixing it with subtraction (19 and XIX), and Mayas organize in yet another way in base twenty. This question irks prehistorians even more: can he say "circle" when he sees a "round" shape that is not oval? Should he try to find a center? But an oval may have a center of symmetry... Would this not mean that for the sake of the realization of the operation, we want the compass to have existed before the circle had been thought of, when the equality of some radii on the drawing may be sufficient? Should we talk of a zero in a written number when nothing is specified and that 1003 and 103 can thus be confused orally?

To come back to the use of the expression "mathematics class", I can only note that it appears (under the pen of Pierre Hérigone[1]) when Euclid's classical book was no longer enforced as the common denominator of all mathematical knowledge.

Indeed, among all sciences, mathematics has the most surprising form, because they are born from a book: Euclid's *Elements* (towards 300 B.C.), which knits together disparate these basic objects you mentioned. In other words, it is not a science in a state of research, but of an established science—somewhat of a small miracle, all the more so as almost nothing is known of its author, as though symbolically his text represented normal human thinking, I do not dare say natural, about basic mathematics. Euclid's *Elements* are a beautiful text, which can still today be read with feeling... Indeed I do not think that reading it can teach us anything, but such a rigorous and organized masterpiece is moving... No other science has such foundations.

Cédric Villani: Euclid's *Elements* are said to be the most published book in mankind's history after the Bible... There is an almost religious aspect behind this piece of work.

Jean Dhombres: A special feature of mathematics is that, for example, the translation of Euclid's *Elements* in Chinese in the 17th century was remarkably influential on the structure of the Chinese language—even if the Chinese find it somewhat hard to admit for reasons that are perfectly understandable! Following this translation, Chinese language started to use two characters to express one thing. In my opinion, this very rare occurrence proves that Euclid's *Elements* touches essential things in human thought. Mathematics of course existed long before Euclid, and all that was to come is not in Euclid. There is nothing on equations, nothing on positional number systems, and nothing anticipating zero for example. Yet, if the decimal existed in China around the time the *Elements* were written, equations appeared several centuries earlier in the Middle East, and the zero as absence of number can be found at the time of Alexandria, then takes really root in India toward the 6th century.

---

[1]Pierre Hérigone (1580–1643) is a Frenchman from the Basque region who, starting from 1634, published a "Course in mathematics" in six volumes (bilingual Latin–French), invented the abbreviation used to express the orthogonality of two straight lines, and we have kept his layout for the so-called "Pascal's" triangle which he explains by using expansions of the powers of a binomial such as $a + b$.

Besides, Euclid's *Elements* are a basic nucleus, not a book on philosophy: it is not a book designed to be thought-provoking—reflection only occurs at a later stage—, it is an instruction book, whose aim is the transmission of knowledge. *Mathemata* is what can be taught, as opposed to what a master teaches us. With some exaggeration, mathematics from this Greek tradition could almost be defined as a science in which the master can be disposed of. I mean that the purpose of mathematics is to do it alone. Cantor's famous expression: "The essence of mathematics is freedom", is appropriate in that doing mathematics is to make them one's own, without taking away anything from anyone. Once one has understood the proof of the so-called Pythagoras theorem on right-angled triangles—this is proposition 47 in book I of Euclid's *Elements*—, external help is no longer needed, and one feels free to the point of considering oneself as coauthor.

Gerhard Heinzmann: I would like to put this into a historical perspective because the objects in Euclid's *Elements* are not the only *mathemata*. For the Greeks, the title *mathemata*, rendered in French by the plural *mathématiques*, denotes the objects taught in the realm of "science," which differs from the realm of experience by the fact that acquisition of knowledge always comes with an argumentative study of the result. According to this definition, one finds not only Euclid's *Elements*, but also philosophy treatises, for example Aristotle's Metaphysics, in the department of "science" because they deal with *mathemata*. But, philosophy or metaphysics deal with *mathemata* in a more general manner, whereas a mathematical standpoint is more limited.

Jean Dhombres: Indeed, I considered that Euclid's *Elements* were the foundation of the specificity of mathematics, which is assuredly somewhat artificial, but historically undoubtedly deep. Because Euclid's *Elements* have erased all mathematical efforts they are derived from: if history remembers some names, Pythagoras, Thales, it does not remember their proofs, and only those kept by Euclid's remains. As a game, let us compare mathematics with metaphysics, which speaks of a certain number of things, organizes a priori arguments, takes shape and multiplies references... As for Euclid, he only refers to what he considers, in other words to a list of actions, of definitions... In the world of mathematics of the 3rd century B.C.—and despite the fact that so much had taken place before, we will regard this date as the starting point—there is a golden rule that can be resumed as follows: "I shall refrain from mentioning anything but what I have defined"—which does not mean that I pretend to know a priori everything that follows from the definition.

Cédric Villani: Anyhow, the starting point is an initial set of rules and care is taken to never use a superfluous rule. So it is with Euclid, but it is not always obvious. For example, an important historical development in mathematics concerns Euclid's fifth postulate, the axiom according to which only one line can be drawn through any point not on a given line parallel to the given line.

Pierre Cartier: This is not how it is formulated; its formulation is far more incomprehensive!

Jean Dhombres: Indeed, Euclid's fifth "demand" (or postulate), at the start of book 1 of *Elements*, occurs as a sequence in a continuous discourse, fixing the framework; it positively announces a meeting, impressively leaving aside the only case of parallelism:

> If a line segment intersects two straight lines forming two interior angles on the same side that sum to less than two right angles, then the two lines, if extended indefinitely, meet on that side on which the angles sum to less than two right angles.[2]

Cédric Villani: Thats true. And in terms of knowledge construction, it is certainly extraordinary to see how generations of mathematicians have tried to get rid of this axiom—and yet there are many axioms in Euclid!—to make things "even more perfect"

Pierre Cartier: Because this one seemed more artificial.

Cédric Villani: Yes, it seemed more artificial, but they found that it is not possible to do without it. And there have been many dramatic changes in its history, since it gave rise to non-Euclidean geometry, and even to general relativity. This almost tyrannical desire to simplify is rather remarkable as in the end the outcome of this type of self-discipline is something full of promise.

One of the characteristics of mathematics is to start from a very narrow base, together with the ambition to describe a great variety of phenomena. In Euclid's *Elements* we already find this contrast between a short list of postulates and a long list of theorems, of situations. The approach is paradoxical: starting from a short list of statements whose truth is assumed, then uses exclusively logical reasoning without introducing any additional axioms—and yet this gives rise to a rich and fascinating world: the geometry of the triangle, Euclidean geometry...

Pierre Cartier: And let us not forget that Euclid did not only write the *Elements*, but also *Optics*...

Jean Dhombres: Quite. This is why I thought it useful to begin by saying that the *Elements* are a hard core which gained in recognition over the centuries. And even if over the centuries adjustments have been made to the text of Euclid's *Elements*, they are comparatively minor. At least until this text was completely discarded from secondary school teaching in the second half of the 20th century. Which raises a question

---

[2]The French translation used is the one given by Bernard Vitrac (Les Éléments d'Euclide, 4 volumes, Paris, PUF, 1990–2001) which we are lucky to have in French and which wisely reviews the tradition of comments on this text, while remaining readable by someone trying to understand the mathematics at stake, and who does not want to get overwhelmed by philological remarks, but merely benefit from his understanding.

that we shall have to address to understand how discontinuities occur in mathematics, and also how this science reacts, and thus understand its deep-rootedness in human history.

Cédric Villani: Moreover, once the importance of the 5th postulate was understood, people came to consider it as evidence of Euclid's true genius, and realized that despite his convoluted formulation this axiom cannot be discarded! Besides the axiomatic debate is still ongoing, debates on the foundations of mathematics are not yet settled, and its internal coherence has never been proved—will it be one day? On a small-scale, one of the battles I led was to get rid of all occurrences of the axiom of choice[3] from analysis proofs in all of my undergraduate and postgraduate courses. To such an extent that in my huge monograph on optimal transport—a subject that may be simplistically summarized as the study of the best goods transport plan between production units and consumption sources: it is a fast developing modern topic, moreover linked to non-Euclidean geometry—I start my book with: "Axioms:... I do not use the Axiom of choice".

Mathematics, a Fact of Civilization

Sylvestre Huet: Let us return to the origin of mathematics, to the construction of mathematical tools and to their relationship with reality, as well as to the interpretation of the most basic mathematical objects. Indeed, prehistorians who look for the oldest evidence of mathematical activity end up with things from several tens of thousands of years ago, for instance evidence of counting. And little by little, counting methods become more and more abstract. Objects are drawn that are initially very basic (a line meaning "1"), then more abstract signs denoting numbers are invented. Since Antiquity, the concepts of points, lines, surfaces appear as mathematical objects without any corresponding physical counterparts in the sense that, if I outline a field for land surveying, I will be drawing a line on a support. Mathematically speaking, this line has no depth. Yet, in concrete terms, no field boundary is merely a mathematical line. How do you picture the intellectual work that needed to be developed to produce these mathematical objects came about? And from that time on, a time deeply influenced by the metaphysics and religion around Pythagoras—even if all Pythagorean writings are not apocryphal—, why did intellectuals who produced these first mathematical objects share a rather Platonic conception of mathematics (that is of mathematics existing in a universe independent from reality)?

Pierre Cartier: To consider this issue, I would like to add another dimension to the discussion: the development of mathematics is a fact of civilization. In other words,

---

[3]The axiom of choice is a seemingly harmless axiom which states that, given a family of sets, it is always possible to choose an element from each of these sets, even if there are infinitely many sets, for as large an infinity as desired.

it should not be seen as an isolated event, but as an integral part of civilizational advancement (which can be made to have started with the invention of agriculture some 10,000 years ago). As soon as society was formed, probably some tens of thousands years ago, rules of negotiation, of conflict regulation, etc. needed to be invented. Civilization can function only with a mathematical investment which seems to us elementary today as it is taught at the very beginning of education, but which is gigantic. As soon as we went beyond barter, money became necessary. As soon as the rudiments of States were organized—it is well known that 6000 years B.C. there were already structured States—, granaries, inventories, accounts, registries were needed... Regulatory mechanisms of both landed or material property require early stages of bureaucracy, investments, and an extensive know-how. We forget this because we are in it, but the mathematization of our environment is astounding, just look around you! Another side to all this is geometry, which arose from architecture. In fact in China, before the introduction of Western science by Jesuits, geometry was not considered as an autonomous science and its vocabulary was that of architecture. The problems of geometric constructions that appear in *the Nine Chapters*[4] in general give the sketch of a city with its walls, its doors, a very detailed map... It is also known that contrary to the Chinese who have plenty of distinctions between geometrical forms, our terminology is not sufficiently precise. To tell the truth, we hardly distinguish between solid geometric forms, to the extent that, for instance, I recently found it difficult to answer the question of a high school teacher who asked me: "What is a cone? Is it the surface or the solid cone?"

Astronomical calendars could also be mentioned... In other words, as societies take shape, mathematical reasoning is drawn upon and develops alongside. Thus, logical reasoning probably developed with law. In all civilizations, progress in abstract reasoning is connected to argumentation—thus the Greeks were consummate polemicists and quibblers!

For me, mathematics is truly a fact of civilization, and this is why the debate pure mathematics/applied mathematics seems to me sterile. Just think of the progress in number systems...

Sylvestre Huet: I would add a political element. The Alexandria school was one of the flagships of ancient mathematics. Behind it, we find a political sponsor, the Ptolemaic dynasty (pharaohs descended from Alexander the Great's Macedonian general Ptolemy, who took over Egypt at the conqueror's death; the last of whom was... Cleopatra). This dynasty, who wanted to dominate international commerce of the time especially in the Mediterranean, needed mathematical tools for navigation, drawing maps, calculating itineraries, etc. To this end, the Ptolemaic state funded the library and the purchase of manuscripts, as well as the reproduction of any manuscript passing through Alexandria. This enabled a concentration—this first cluster!—of engineers, mathematicians, shipbuilders and individuals capable of organizing complex economic activities...

---

[4]Chinese mathematics book from the 2nd–1st century B.C.

Jean Dhombres: We could even go further back into the history of mathematics, at the time of prehistoric man, with an incredible example which takes us back to the birth of abstraction. Indeed we find a difference in the processing of two-sided flint tools: concerning the organization of sketched figures, some stonemasons conceived the tool as a solid, in other words as a three-dimensional object, whereas others conceived it as a two-sided plane, in other words as a two-dimensional object. By contrast, it cannot be said that the third dimension succeeded the second dimension, and the two types are seen to have coexisted, like distinct families. Hence one can already talk of some form of "civilization," since to the construction of two-sided tools already requires the implementation of a small industry and abstraction since the drawing on the finished object varied according to the way the mason perceived the form of the object.

Pierre Cartier: This is certainly one of the major steps.

Jean Dhombres: Hence there was indeed a distinction between the plane and the space, and this is definitively observed. We who are onlookers can classify. On the other hand, it is impossible to know whether prehistoric men were aware of this distinction between a two-dimensional two-sided tool and a three-dimensional one. One can in particular refer to Olivier Keller's[5] work which raises the question of abstraction. At least, mathematics in its initial stage did not follow the Cartesian order, namely it did not necessarily go from the simple to the complicated.

---

[5]Olivier Keller, *Aux origines de la géométrie, Le Paléolithique et le monde des chasseurs-cuilleurs*, Paris, Vuibert 2004.

Mathematics and Reality

Mathematics as a Tool to Understand and Act on Reality

Sylvestre Huet: To start this chapter on the nature of mathematics and of its relationship with reality, I would like to quote you two extracts from interviews I carried out for *Libération*,[1] one of a mathematician you are familiar with—Alain Connes—and the other of a Belgian theoretical physicist and epistemologist—Dominique Lambert. Here is what Alain Connes said:

> There are two opposing extreme viewpoints about mathematical activity. The first one, which I am entirely in agreement with, follows the Platonists: it states that there is a raw, primitive mathematical reality which predates its discovery. A world whose exploration requires the creation of tools, in the same way as ships had to be invented to cross oceans. Mathematicians will, therefore, invent, create theories whose purpose is to shed some light on this preexisting reality. The second viewpoint is that of formalists; they deny mathematics any preexistence, considering that it is a formal game, founded on axioms and logical deductions, hence a purely human creation. This viewpoint seems more natural to non-mathematicians, who are reluctant to assume an unknown world which they do not perceive. People understand that mathematics is a language, but not that it is an external reality outside the human mind. Yet, the great discoveries of the 20th century, in particular Godel's work, have shown that the formalist viewpoint is unsustainable. Whatever be the means of exploration, the formal system used, there will always be mathematical truths beyond it, and mathematical reality cannot be reduced to the logical consequences of a formal system.

In contrast to this vision of mathematics, Dominique Lambert expressed the following point of view in another interview published in *Libération*, attempting to resolve the difficult issue of the relationship between mathematics and reality by a historical approach:

> "A mathematician seemingly invents with imagination as his only guide and mathematical rules as the only rules. Thirty years later, his invention helps to describe a particle or space-time. Why? Mathematics are efficient. Very much so. To the point of causing turmoil among physicists, the biggest "clients" of mathematics. Princeton physicist Eugene Wigner's

---

[1] A French daily, translator's note.

© Springer India 2016
P. Cartier et al., *Freedom in Mathematics*,
DOI 10.1007/978-81-322-2788-5_2

famous article is an example. He called it: *"The Unreasonable Effectiveness of Mathematics in the Natural Sciences"*.[2] A compelling formula, but an old problem. As far back as Antiquity, numbers have been assigned some sort of power. Pythagoreans saw the meaning of the world, its true reality, in numbers to which they assigned a metaphysical status. With Plato, mathematics became a world apart, separated from the world of appearances, which provides access to that of ideas. As for Aristotle, he saw mathematics as a mere abstraction based on reality. And the debate is unable to find a way out of this brutal dichotomy. Each school accumulates examples supporting its case without being able to oust its opponent who opposes as many counterexamples. To break this deadlock, one needs to observe the history of the relationship between mathematics and physics. It is then possible to justify both these conceptions and to nuance their element of truth. The great rise of the Platonic conception goes back to the birth of modern physics. When Galileo wrote: "The great book of Nature is written in mathematical language", he was not surprised by the efficiency of mathematics, he assumes it. According to him, it is necessary to understand and to express the world through mathematics because mathematics has something to do with the structure, the essence of the world. And he is believable because his approach is efficient. Forgetting that he carefully chose basic physical systems (fall of a marble on the floor) that allow for approximations and simplifications. We forget that the mathematical notions he used—the point (without extension), the line (without depth)—have no physical counterparts. However, there was subsequently a tremendous upsurge in the development of mathematics. It often appeared to be independent of all else. This gradually reinforced this conception, which reached its peak with Cantor's sentence: "The essence of mathematics lies in its freedom". And with Bourbaki—un group of French mathematicians formed in the 1930s—for whom mathematics is a self-sufficient whole. This conception gives a "meta-physical" dimension to mathematics and separates them from empirical science and the world. In this context, efficiency can only be unreasonable, even miraculous. The "result of a pre-established harmony", as stated by Leibniz. This position is all the more powerful as the efficiency of mathematics in natural sciences, biology, and sometimes in economy cannot be denied. But is this efficiency connected to a "metaphysical" structure of the world? Or to some other reason, which is in no way miraculous or mysterious? To answer these questions, it is necessary to to delve into the history of mathematics. We then find that it looks like a co-evolution with natural sciences and especially with physics. Think about the co-evolution of flowers and pollen-gathering insects. Except that there are phases of relatively independent development as well as periods of strong interactions with empirical domains. The latter bring new problems, new information to mathematics which takes hold of them to initiate a new development of its own. Some sort of bilateral self-stimulating interaction, with phases of independent developments. But a co-evolution where the main stages of each protagonist do not necessarily coincide, contrary to the biological image. There are decades, if not centuries without any interaction. After Riemann developed curved spaces around 1850—this is pure geometry—six decades elapsed before physics grabbed hold of it with Einstein's general relativity.

Because of this time lag, isolating one of these sentences, you find a mathematician who asks himself questions for their own sake, and very rich domains appear to develop by themselves. Alain Connes described them as "generative domains", Jean Dieudonné of "fertile problems", which seems to agree with Platonism. But if we broaden our outlook, we see that, historically, mathematics would not have alone reached its present development. How can the development of integral calculus or of analysis be imagined without mechanics? It is because Werner Heisenberg strove to "glue" to the spectrum of the hydrogen atom that he managed to build the first mathematical apparatus of quantum theory. Later, Richard Feynman's work in quantum mechanic led to a whole series of mathematical developments. In the 1970 and 1980s, chaos theory and dynamical systems were stimulated by meteorol-

---

[2]E.P. Wigner, *Communications on Pure and Applied Mathematics*, XIII (1960), 1–14.

ogy or the desire to understand turbulence. If mathematicians were left alone too long, they would not go very far. Platonism does not account for this. Aristotle's shortcoming was to believe that mathematics is merely derived from reality. Plato's shortcoming was to believe that all mathematics exists in a universe autonomous from reality.

The connection between mathematics and reality proves to be more subtle. Let us start from vision. I see a cup and consider it real because when I look at it from different angles, its perception remains the same. If changing angles suffices to change the cup, it is an illusion. Behind the idea of invariance is the idea of reality. The physicist does the same. He recuperates sets of objects, defines operations on these objects, then, physical dimensions and measures are associated to concrete invariants: momentum, energy, charge. However, is this not precisely what mathematicians do: define sets of objects on which operations are carried out and where invariants are searched for? Mathematics is some sort of extension of the process of perception.

This close proximity between mathematics, physics and perception, explains why the formal nets woven by mathematicians can catch bits of reality. Conversely, the relative independence of mathematics explains why some of these nets do not catch anything. But also why several different nets may describe the same reality. Moreover, that the same equation can describe— as shown by Jean-Marc Lévy-Leblond—different natural phenomena shows that it does not touch its essence, even if this equation is often essential to think about them. The equation of the electron is not the electron. As the great mathematician René Thom warned us 'describing is not understanding'.

Gerhard Heinzmann: These interviews show that the attempt to describe the relationship between mathematics and reality by having recourse to the Platonic or the formalist position has shortcomings and leads to philosophical deadlocks which, firstly, do not have easy answers and secondly, cannot be broken by merely observing the history of the relationship between mathematics and physics. So as not to misuse the time allocated to me, I will omit the formalism (which, contrary to what Alain Connes said, is always sustainable[3] and I will confine myself to a brief discussion of Platonism: it is one thing to assume a primitive mathematical reality preexisting its discovery (every mathematician is free to believe in it and he may find it quite useful to do so) and another to give rational arguments explaining how these abstract entities can be grasped (these arguments are the task of the philosopher) without falling into the trap of an analogy with the perception of concrete objects. Thus, to avoid presupposing the mysterious capacity that gives access to Platonic entities, following the Duhem–Quine thesis,[4] philosophers have defended a moderate Platonism: quantifiers of a theory relate to sets of numbers, to functions as well as particles, and fields. It, therefore, seems reasonable to believe in the existence of theoretical elements in the same way as we believe in empirical elements as long as the theory resists critical tests. Nonetheless, this standpoint leads to a number of questions[5] and

---

[3] See for example Michel Detlefsen (2004), Formalism, in: Stewart Shapiro (ed.), *The Oxford Handbook of Philosophy of mathematics and Logic*, Oxford, Oxford University Press, pp. 236–317.

[4] This thesis states that it is impossible to test empirically a mathematical or physical hypothesis in isolation, but a bundle of physical and mathematical hypotheses can be tested.

[5] What are the mathematical entities truly vital for a given scientific theory? If it could be shown that these are numbers, the Platonic answer could reduce to a constructivist position! Which principles concerning these entities are necessary for the required mathematics?.

requires paying a heavy price: a clear semantic separation between abstract objects and concrete objects must be refuted.

Defining like Dominique Lambert mathematics as "some sort of extension of the process of perception," nothing is gained philosophically, and one remains stuck in the analogy.

So I propose another solution. It consists in saying that mathematics is a language used in the sciences to *decomplexify* reality—which may seem surprising to someone who has not done mathematics, and who finds it extremely complex. But in the end it is reality that is complex. And mathematicians, from the beginning, have been inventing a language to measure it, to simplify it, to decomplexify it. This is what Euclid did in geometry. In arithmetic, one needs to wait till Peano[6] for structures to be highlighted and axiomatic systems elaborated. Hence, arithmetic is seemingly more complex than the structure of geometry, which is also paradoxical. Moving away from numbers accessible to daily experience and writing down their structure was harder than moving away from concrete drawings.

This is what is essential in mathematics: it is a science which reflects on necessary structures to express the complex reality whose objects are "invented" as a position within a structure.[7] It does not describe reality, but serves as a tool to express it. Mathematics is our tool to access a reality, generally speaking, more complex than the one accessible to our senses and our experience.

Cédric Villani: I go along with what Gerhard just said, especially about the fact that mathematics initially arises from a desire to simplify. Indeed, what surrounds us is complicated, incomprehensible, and besides, what is reality? Even the simplest things.... Phase transition is a perfect illustration of this: take a pan, boil water, and what happens? Water goes through a phase change… and nobody has ever understood why! This is one of the still open problems in mathematical physics. The complexity that surrounds us is absolutely terrifying if you try to imagine it. The mathematical approach consists in extracting guiding principles that will enable us to describe the world around us, then to understand this world, and finally to act upon it. To understand and to act. These are the two basic motivations of the mathematical approach: the one is inseparable from the other—and conversely. To this end, all lines, all paths, are simplified and replaced by a line, some sort of extremely simple representation; for somewhat more complex forms, there is the

---

[6]Giuseppe Peano (1858–1932), Italian mathematician and linguist. A pioneer in the formalist approach to mathematics, he contributed to the development of an axiomatization of arithmetic.

[7]By using the term "invented", I obviously do not mean that the mathematician invents material objects, but that the mathematical conceptualization of reality is not a description, but preserves a certain degree of freedom. To illustrate what I want to say, it is convenient to take the traditional example of geometry before 1905 (space-time); I can either take Euclidean geometry or one of the non-Euclidean geometries to measure street angles; none of the descriptions obtained is more true than the others and for this reason none is a *description* in the sense of the expression of a biunivocal relation, but is an "invention". Distinct theories are, therefore, empirically equivalent or, as formulated by American philosopher Willard Ban Orman Quine: theories are sub-determined by experience.

triangle… And gradually, one comes closer to the complexity of the world around us, by attempting to rely on a set of simple axioms, simple rules, logical rules that enable us to somewhat find our bearings, and put some order in a world so indecipherable that it is absolutely brutal.

And coming back to the initial question on abstract/concrete aspects, it is a dialectic approach from the outset: going toward the world, returning, looking, reflecting. This differs from an experimental process (reflecting on an experiment is followed by testing it against the world, and then returning), but still it is a bit like that: a theory is elaborated, observation shows whether it provides a good description, viewpoints are changed, and so on. Take the parabola, which is one of Galileo's greatest successes: he made a flagrant mistake, and what is remarkable is to see him finally reach a correct result (in other words, the trajectory of a parabola) by accumulating mathematical mistakes. It may seem simple today, but at the time finding the trajectory of a projectile was extremely complicated. People continued to get excited by this for a long time,

Pierre Cartier: Moreover, for Galileo, the continuous motion of a solid (according to the inertia principle he himself defined) is not a motion along a straight line at constant velocity, but a circular motion at constant velocity! Nonetheless, if the parabola is identified to its osculating circle,[8] in the conditions of the experience, we recover something coherent.

Cédric Villani: This example is a good illustration of the extent to which what surrounds us is complicated. In Galileo's times, the arts were flourishing, there were already masterpieces in literature, painting, sculpture… but in terms of the description of the familiar world around us by natural laws, reflection still centered around trivial issues such as: "If I throw this object in the air, how is it going to fall down?"

Jean Dhombres: Let us then play an intellectual game to place our discussion in the thinking of an earlier period… I am professor of Aristotelian physics in Padua around 1604, and so am a colleague of Galileo and I also teach about falling bodies— for such is required by the program which enforces a commentary of "*Mechanical Questions*", a book then attributed to Aristotle and dealing with heavy bodies thrown through the air. Now, my colleague Galileo uses simplification of a mathematical nature (the fall of bodies in the void is seemingly founded on a new law expressed mathematically), whereas this is a real phenomenon. Galileo can only be wrong for scrimping on Aristotelian thought like a bad student and by denying all efforts to interpret the texts of so many past professors. In the name of academic dignity itself, I propose a compromise theory, which includes various positions, one of them being taken from a book written more that 70 years ago by a mathematician named

---

[8]The osculating circle of a curve at a point is the circle best approximating the curve passing through the point, keeping its direction and curvature.

Tartaglia,[9] who introduced so many novelties in the algebraic domain of equations. As an accredited professor, I dogmatically stipulate that when I throw a stone in the air at an angle, the stone first goes straight because when throwing it, I gave it this *impetus* which I refrain from defining or quantifying; but this *impetus* runs out and when it has run out, at the end of the straight line, the stone no longer knows where to go. Not in mere rhetoric can I say that not knowing where to go, it goes in a circle. When it returns to the same height, the stone lets itself fall straight down. For those of us reading today, it is clear that there is not *one* trajectory, but a variety of motions indicated by different forms, curves or lines. There is a rectilinear motion, then something emotional, or psychological: the stone thinks, it thinks it does not know where to go, hence it goes in a circle. Then it falls straight down because it tries to reach the center of the world to which it aspires, since it has taken a fancy for this imagined center.

Cédric Villani: One could play devil's advocate and say that these arguments appear to be much richer and more interesting that Galileo's spartan arguments.

Jean Dhombres: Precisely. There is something rude in the mathematics presented by Galileo. Does not the latter say one should forget "fancies"? When his frequently polemical texts are read honestly, despite the experimental and deep side about the parabola which is the only curve he finds for the trajectory of the stone thrown,—we have drawings and accounts of fairly brilliant experiments by Galileo concerning the parabola—, we basically get the impression that the experiments come later. After coming up with the idea that there is no dissipated *impetus*, the unique trajectory stems from the composition of two totally distinct effects and that are both standpoints on our perception of the world. "Normally" the trajectory should be "a straight line with uniform velocity". This uniformity, which is the exact opposite of the dissipation of the *impetus* because the fact that Galileo enforces the conservation of velocity means that he represents time on a line, and that he makes equal amounts of distance covered correspond to equal amounts of time. But at every position that the stone should occupy on the line followed when thrown, the stone falls, this is the second effect. Galileo checked that the length of vertical fall is proportional to the square of the time elapsed since the start of the throw—such is the aim of experiments on the motion of a heavy ball on an inclined plane representing an artificial decrease in velocity of the vertical fall giving rise to observation by bringing into play the slope parameter of the inclined plane. Which means, and here he uses mathematics that then seemed sophisticated, that the length of the vertical fall at each point is proportional to the square of the inclined distance. As any high school student today would recognize, in one go I obtain the Cartesian equation of the parabola, with respect to non-orthogonal axes, with on the one hand the direction of the throw and

---

[9]Niccolo Tartaglia (around 1500–1557) was an Italian mathematician, who gave away his technique for finding the roots of polynomial equations of the third degree to his fellow countryman Girolamo Cardano (1501–1571), Cardan in French, translated Euclid into Italian, and among other domains worked in ballistics.

on the other the vertical at the point of throw. This amounts to analytic geometry. On no account is it a proof in the physical sense, and Galileo presents arguments in which the terms used fit a mathematically formulated definition, contrary to what was done with the *impetus*. The role of experimentation remains modest here; it is a question of constructing mathematical facts, uniform velocity and vertical fall; simplification lies in the exclusion of all the rest, that is to say of the form of the stone thrown, air resistance, the place where the stone is thrown, etc. Today we would say that Galileo has eliminated all "noise" around the phenomenon. And two physical concepts then arise: initial velocity, which happily replaces *impetus*, and acceleration, which we call gravity. Economy, simplification and indeed, one can speak of an abstraction of reality, and even of a rift, in the sense "I kill reality" to conceptualize it. Mathematics plays with reality, with the purpose both of describing in the simplest manner and of being able to act, in the manner of an experiment. Because I am not at all sure that the purpose is—at least for Galileo—to understand a priori. To use scholastic language, Galileo did not search for the substance of the fall, but its form, and mathematical abstraction enabled him to succeed physically, and led him naturally to ballistics and water jet techniques.

Pierre Cartier: Indeed Galileo was foremost an engineer; he built fountains…

Jean Dhombres: I was almost going to say that his a priori mathematical understanding of the fall of bodies is indissolubly linked to his physical sense. Let me add that since Galileo, no one has seen a water jet coming out at an angle from a fountain in some park in the world without detecting a parabola. This realization of mathematics by a curve invented so many centuries ago is for me one of the most enlightening signs of modernity. Because, it indicates a modification of Greek mathematics, in the sense that one recovers geometric properties of the parabola (such as the existence of an axis of symmetry, whose perception is not obvious at the outset when one looks at a water jet) only because one gives an analytic description of it which here takes the form of a trajectory—what specialists call a parametrization. The object "parabola" has, therefore, changed even in the eyes of mathematicians: it has become enriched with an interpretative network. Similarly human consciousness, which out of the parabolic form of water jets created symmetry in French style gardens, Versailles being one of the most successful examples.

Cédric Villani: There are two trends: the "we want to understand" one and the "engineer" one. And both of these trends co-exist. We find them in Galileo and others. Let us mention Euler, who started fluid mechanics because he too had fountains to build and all sorts of things.

Pierre Cartier: Moreover, the first prize obtained by Euler from the French Academy of Sciences concerns boats and sails, when he had never seen a single sailing boat— not many could be seen in Basel!

Cédric Villani: Let us also mention Newton,[10] who wanted to understand the order governing the planets. Here we have a desire to understand, with also a rather surprising relationship between reality and theory. It is said that Newton somewhat altered the results of his observations so as to make them fit well with his theory.

Pierre Cartier: Copernicus and Kepler as well. They all cheated a bit!

Jean Dhombres: Certainly! But did they cheat more that the Padua professor in my story, or more than those who wanted to keep the Ptolemaic system?

Mathematics as a Self-Sufficient Whole

Multiple Realities, One Language

Sylvestre Huet: In other words, mathematics is developed primarily to describe the world, then to understand it, and as the world is complex, it is simplified. Moreover, at the beginning, the most simplifiable natural phenomena were chosen so as to be able to use mathematical objects and mathematical methods available in earlier times. Besides, from then on, the mathematical part of all this becomes self-contained in the sense that there is already a reflection about the most basic objects intending to go beyond a simple abstraction of reality in order to simplify it, describe it, understand it, even act on it. Why, how, and what are, in your opinion. the consequences of this mathematical development which, starting from this period, desired to cut the link with the reality of the physical world—even agree with the Platonic vision: the meaning of reality would seem to be in this abstract definition and not in the reality itself. It is Plato's cave: the idea of the table is better than the table, the latter being only a more or less perfect realization of this idea. Is it still thought today that nature is written in mathematical language (Galileo) or that there is a "pre-established harmony" (Leibniz) between mathematics and reality?

Jean Dhombres: I will go back to Galileo's example. Basic objects held a fascination for the mathematician Galileo. And when he discovered with the help of his parabola, which is not a basic object, that he can say that the distances covered in equal amounts of time are like odd integers (1, 3, 5, 7, 9…), he was fascinated by these ancient objects that are not arbitrary numbers. A study that may be considered a study of reality, even if many things have been set aside since only the stone's motion is being observed,

---

[10]Isaac Newton (1642–1727), English mathematician, physicist, philosopher and astronomer. This emblematic scientific figure is especially known for having founded classical mechanics, elaborated the theory of universal gravitation and, along with Leibniz, developed infinitesimal calculus.

now included objects, odd numbers, that have been already studied and that have well-known properties...

Pierre Cartier: To understand the sudden appearance of odd integers here, the *gnomon* needs to be mentioned. This plane figure known since time immemorial is a square made up of small squares and cut into strips. The figure readily shows that the strips are successively made up of 1, 3, 5... small squares, thus implying relations:

$$1^2 = 1$$
$$2^2 = 1 + 3$$
$$3^2 = 1 + 3 + 5 \ldots$$

which are the ones that fascinated Galileo.

Jean Dhombres: Exactly, and when Galileo found this correspondence, he marveled that seemingly independent things, namely number games, appear at the core of something real. To recover something which seems to be from "elsewhere" starting from what can be labeled some form of reality is almost a proof of truth for Galileo. It cannot be a matter of chance. However, the proof is not of mathematical type. I should, therefore, like to stress that Galileo's famous 1623 book "The Assayer" does not mention mathematics as the language of nature, which at the time meant the language of physics in the old sense, but as the language of the universe. Probably not so much to describe all there is, and refer to reality, but to encompass all that has a structure, and in particular he could only be thinking of the Solar System.

Gerhard Heinzmann: I think you will find an answer to your question if you change a word in your description. It suffices to say that mathematics helps to *articulate* reality, instead of saying that it *describes* reality. If mathematics is considered a language, namely that of the most general relations, the mathematician takes "his" language as the object of his study, just like the grammarian looks at the structure of the natural language. However, mathematics can be reduced to logic and to set theory (belonging is then the only relation related to a content). This leads to more or less two possibilities: either you consider sets as primitive abstract objects (in a Platonic perspective or an instrumentalist perspective) forming the positions of your mathematical structure (=language), or you consider the structure of the sets themselves as a primitive object. It is then a useful tool but in principle open to review.

I think it is better to get free of the idea of Platonism with respect to objects—there are mathematical objects in a self-contained universe with respect to reality... I am not a historian and I do not consider the real development of mathematics, but I am interested in your question on the language aspect of mathematics from a logical point of view. And from a logical point of view, there is no difficulty in explaining that a mathematician is someone who develops a language he understands. We all speak a language, we all make abstractions using plain language: for example, if I show you a table saying "table", it is an abstraction for you, you cannot even really see this table as it can be seen from so many possible angles... And yet, the word "table" is as a general rule sufficient to very sensibly indicate the object singled out, though the object is not strictly speaking defined. Primitive mathematical objects (like structures or sets) are obviously very abstract objects but sufficiently sensible to be used as the foundation of a theory. The language of mathematics is "in harmony" with reality because it is our tool for getting to know (using appropriate scientific definitions) reality and not because nature is written in mathematical language or because there is any preestablished harmony.

Mathematical Tools to Articulate Reality

Pierre Cartier: I am going to incorporate a second keyword in our discussion. It is *tool*. On the one hand, language and on the other, the tool. I am extremely struck by the mastery of the mathematical tool acquired by ordinary people in the last 50 years. Allow me to mention an anecdotal example: when I was 14, at the end of the war, a mutualist school was founded in my hometown on Proudhon or Bertrand Russell model. The purpose was to educate young workers to enable them to obtain an advanced specialized technical degree or its equivalent. Facing me were 18-year-old "kids"—when I was only 14—and I used to teach them mathematics. The lesson on negative numbers set me the greatest difficulty. It was really stupendously difficult to explain that minus multiplied by minus gives a plus, etc. Sixty years later, I had left my car in an underground parking and asked my 4-year-old granddaughter whether she remembered where it was and she confidently replied; "in -7". At 4, my granddaughter knows what a negative number is, she feels it, the tool has been incorporated.

Let me take another similar example: my son-in-law teaches in a technical high school. One assumes that in a technical high school, the level of mathematics is surely not very high. Well this is wrong! His mathematics class is more difficult than what he would teach in a general high school, but it does not focus on the same things. For instance, he teaches complex numbers[11] in his class on electricity in conjunction with experiments. Yet, I can assure you that in my final high school year, our mathematics teacher did not know what a complex number was. Our physics teacher was slightly more knowledgeable. Today, it is part of standard skills, of our background. To some extent I think of the progress of civilization as the gradual expansion of the notion of a number, and more precisely of the notion of numbers that can be manipulated. Obviously today, the notation 10^x has come into common use: there are megas (10^6), gigas (10^9), teras (19^{12})…, which was not the case forty years ago.

---

[11] See note p. 52.

On the other hand, I also insist on the fact that a new tool is widely used since the 20th century (invented in the previous century): the *matrix*.[12] When Heisenberg[13] gave his matrix formulation of quantum mechanics, he went to see Born[14]—who was his boss—and the latter said: 'Wait,this reminds me of something! There is a construction resembling this in a former mechanics course. Lets see this together"! Hence in the 1920–1930s, the best minds discovered matrices—I must admit that this is not quite true, some algebraists (Jordan, Minkowski, Artin...) had already manipulated them, but more confidentially. However, matrices have now become a basic tool, in particular in statistics. Likewise, even at a basic level, electrical (impedance, etc.) and linear optical laws are no longer taught without mentioning that there are matrices behind. A good course must mention them. Furthermore, it is a usable tool, you can compute eigenvalues on your calculator, etc. Now, let us not forget that most of Le Verrier's[15] work on the discovery of Neptune is based on the diagonalization of 4 by 4 symmetric matrices. Today with a calculator he would get the result instantaneously.

Mathematics thus create tools, and society's appropriation of these tools is part of mathematical progress. It can, therefore, be said that in some simplified way, there is an investment starting from reality, an abstraction, and a return to reality.

Cédric Villani: The notion of tool described by Pierre is central, and to come back to the question about why one studies mathematics "in themselves", sometimes in a disconnected manner from the "real world", I would say that in mankind's history, each time a new tool appears, in any field, there is always someone to rush in and study for its own sake. There are specialists in political law for the sake of political

---

[12] A matrix is a square or rectangular array of numbers, such as

$$\begin{pmatrix} 0 & 1 \\ -1 & 0 \end{pmatrix} \text{ or } \begin{pmatrix} 2 & 3 & 4 \\ 9 & 8 & 7 \end{pmatrix}.$$

Two matrices of the same size (2×2 for example) can be added by adding the corresponding entries together:

$$\begin{pmatrix} 7 & 5 \\ 4 & 3 \end{pmatrix} + \begin{pmatrix} 2 & 8 \\ 9 & 11 \end{pmatrix} = \begin{pmatrix} 7+2 & 5+8 \\ 4+9 & 3+11 \end{pmatrix}$$

There is also a multiplication rule. Let's just say that:

$$\begin{pmatrix} 0 & 1 \\ -1 & 0 \end{pmatrix} \times \begin{pmatrix} 0 & 1 \\ -1 & 0 \end{pmatrix} = \begin{pmatrix} -1 & 0 \\ 0 & -1 \end{pmatrix} \text{ so that the matrix } \begin{pmatrix} a & b \\ -b & a \end{pmatrix}$$

is a faithful representation of the complex number $a + bi$. As a result, calculation rules for complex numbers are justified by matrix calculus. In geometry, a rotation on the plane (or space) is represented by a 2×2 (or 3×3) matrix and matrix multiplication corresponds to the composition of rotations.

[13] Werner Heisenberg (1901–1976), German physicist, who received the Nobel prize in physics in 1932, famous for his uncertainty principle.

[14] Max Born (1882–1970) German theoretical physicist, who received the physics Nobel prize in 1954 for his remarkable work in quantum theory.

[15] Urbain Le Verrier (1811–1877), French astronomer and mathematician specialized in celestial mechanics, discoverer of Neptune and founder of modern French meteorology.

law, specialists in economy for the sake of economy, specialists in the dialect of the Isle of Man or in I do not what, who are in this field *for its own sake*, who go more thoroughly into it *for its own sake* before interacting with others. Likewise, we mathematicians have an extraordinary tool, which we study for its own sake, in all its aspects, which we take to greater depth and so on.

I would like to add two remarks on the question of status. At first, we talked about simplification starting from the real world, in which the dual motion of enrichment and explanation which exists between reality and mathematics is already underway. From time to time, the need to explain something arises and enriches the existing corpus with a new mathematical concept which takes its place as a new tool. For personal reasons, I found one of these concepts fascinating. It is the notion of *entropy*,[16] introduced in statistical physics by physicists Boltzmann[17] and Maxwell.[18] I want to emphasize that Boltzmann created a mathematical concept to understand the physical world. Once this concept became an object, a mathematical object, this tool was further developed and used as mathematics in numerous fields. In a single motion "from the concrete to the abstract and then back to the concrete", entropy can be found everywhere, in every form, in problems concerning the study of gases as well as information exchanges between mobile phones, or even the study of living phenomena… We see how a new tool is shaped: we have a new concept, a new mathematical tool, which is first studied as such—entropy for entropy's sake—, then come those who make the link, the bridge. There is a constant dialectic, concrete/abstract, mathematics/return to the real world.

Complex Numbers: A Tool Then a Reality

Jean Dhombres: In my view, this idea of construction based on a back-and-forth approach between phenomena and concepts is essential to understand the role of mathematical tools even nowadays and since the times of Galileo, Kepler, and Descartes. So I would l like to give another historical example, which always perturbed me, that of complex numbers.[19] They are known to have been developed in

---

[16]Entropy is a quantity characterizing the disorder of a system.

[17]Ludwig Boltzmann (1844–1906), Austrian physicist, founder of statistical mechanics.

[18]James Clerk Maxwell (1831–1879), Scottish physicist and mathematician, famous for his equations unifying electricity and magnetism, as well as for his work on the kinetic theory of gases.

[19]A complex number: it is known that the square of an arbitrary number (integer or decimal) is positive, so -1 cannot be a square. However, if we wish all second-degree equations, without exception, to have two roots, then admittedly, the equation $x^2 + 1$ has two roots, i and $-i$, using conventional notation. If we combine the new number i with the old ones, and want to perform all the usual operations (addition, subtraction, multiplication, division), all these numbers can be reduced to the normal form $a + bi$, which defines complex numbers (as opposed to usual numbers such as a and b said to be "real"). Gauss, Argand, and Cauchy's great discovery around 1800 was to give a geometric representation of these numbers: $a + bi$ corresponds to a point on the plane with coordinates $(a; b)$. As Euler found out, trigonometry can be very much simplified using complex numbers.

the 16th century as a pure mathematicians' object to solve polynomial equations of the third and fourth degrees—one hardly used to go beyond. These numbers play no role in what can be alled the scientific revolution, even if Descartes devised the far more general idea of imaginary numbers. And then, almost one and half century later, after some calculatory type work like that of Viète on trigonometric formulas expressing the cosine or the sine of the multiple of an angle as a polynomial function of the same cosine and sine of this angle, complex numbers appeared in a completely different form. As the complex exponential presented by Euler in 1748 in a book with a fascinating title, *Introduction to Analysis of the Infinite*—a title unfortunately reduced in French to "infinitely small". This second rebirth is the real birth of complex numbers, though still called "imaginary quantities", following Descartes. Euler obtained his famous relation $e^{ix} = \cos x + i \sin x$, which connects complex numbers to reality in that the geometrical angle, here, x is henceforth precisely represented by a purely analytic formula, and in which what is denoted by i here is represented by Euler as an expression whose square is -1.

Pierre Cartier: Indeed, this is how the theory of angles was really founded.

Jean Dhombres: And Euler realized this immediately and formulated it as: the complex exponential underpins the angle.

Pierre Cartier: The same holds for trigonometry, which had no logical foundation before.

Cédric Villani: It also underpins the construction of π. Until the 20th century, the definition of π remained a matter of controversy, to such an extent that the German mathematician, Edmund Landau, defined π as follows: after constructing the complex exponential, π is defined as the smallest root of sine. Unfortunately for him, in his days, his work was considered anti-Aryan or at least degenerate, because having such a twisted mind... Euler's relation was still controversial two centuries after its discovery.

Jean Dhombres: That is quite understandable: it was an analytic construction (let us say of an argument not focused on a figure) which shed light on (I do not say for a moment contradicted!) what the geometry resulting from Euclid had not been able to fully master. But once Euler obtained this construction, a change becomes visible in the mentality of mathematicians: they then tried to find properties of complex numbers for their own sake. We would like to say that they identified two structures: a two-dimensional normed real vector space structure (what is called the topological plane), and a field structure, to use terminology in vogue for almost two centuries, even that they obtained the only commutative field structure not consisting of real numbers, but this will only be proved by the Russian school in the 20th century, in the theory of Banach algebras. To put it differently, we would like these 17th century mathematicians to have interpreted as necessary the operations of addition and multiplication of complex numbers regarded earlier as mere calculation methods.

However, the addition of two complex numbers can be seen in the form of a parallelogram whose construction is based on two of these numbers starting from the origin and whose diagonal is the sum.

The second structure, that of a field, reduces the question to that of the angle of geometry, understood as similitude: it suffices to see that multiplying by Euler's e^{ix} is the same as performing a rotation of angle x. Taking the norm of a vector, also called the absolute value of the complex number, multiplication will then give a homothety. Homothety and rotation, by composition we get all planar similitudes. But this is not what happened: even among brilliant mathematicians, the idea spread that Euler's formula justifies the unproblematic generalization of all mathematical formulas to complex numbers. It is Cauchy who, through his 1821 course, managed to put an end to blindly optimistic analytic practices, which in effect were destroying the entire mathematical tradition.

In my view, what is surprising is that 50 years earlier, mathematicians, in particular D'Alembert, had decided to study axiomatically matters concerning the angle, especially to find the minimal assumptions on which it is possible to found addition of unnamed objects that had become familiar in the form of force because of analytic mechanics and that today could be called vectors. Remarkable work followed in so far as addition which we call vectorial was found to be unique. But they do not put forward the vector space structure. The interpretation of the tool "complex numbers as elements of a two-dimensional normed vector space" had to go through another event: the geometric interpretation of complex numbers. This is a major reform of Euclid's geometry because it provides a means to compute punctual planar transformations. The introduction of this fundamental representation—far from minor—is mostly due to Argand,[20] considered a "minor author". He understood something and successfully imposed it (but how slowly…) because with his representation, he could prove—in less than a page—that every polynomial has at least a complex root. Even though a topological property was missing, it was clean, new, in line with the rigorous spirit of the times. The advantage of Argand's approach is that it grasps what it is that makes the difference between real numbers, over which it is only possible to go straight in one or the other direction around a point, whereas for complex numbers, one can go round a number, the operation being performed algebraically by multiplication, and hence by operations well adapted to polynomials.

Cédric Villani: Is it then Argand who is at the origin of the fundamental theorem of Algebra?

Jean Dhombres: Yes, in the sense that he was the first to give the simplest proof, which is today adopted by every textbook, but no one wants to recognize this! In France it is in general attributed to Cauchy, who only copied Argand, hiding the fact he had borrowed it.

---

[20] Jean-Robert Argand (1768–1822), Swiss mathematician, who worked in Paris and was known for having introduced the planar representation of complex numbers in 1806.

Pierre Cartier: Gauss, like Cauchy, also copied, but these are illustrious names remembered by the history of mathematics, contrary to that of Argand. No one reproduces the frightfully complicated proof given by Gauss in his thesis in 1799.

Cédric Villani: This theorem is even called "D'Alembert's fundamental theorem of Algebra".

Jean Dhombres: Except in one German tradition, where it is still called Argand's theorem—and clearly this is a tradition that goes against Gauss! My purpose was not to take a historical standpoint, but to underline that a purely constructivist theory—"I take elements and then I build"—does not always work.

Cédric Villani: You are a bit harsh when you say that the constructivist theory does not always work, but let us say that indeed it does not lead to everything. It is a part of the structure and then there are also bridges, motivations...

Jean Dhombres: Exactly. To be precise, I think there is in Argand what I would call intuition with all due precaution, as it is the growing awareness of the reality of what "going around" means, which can be made algebraic in that it can be expressed in terms of polynomials. The precise statement of Argand's theorem that "in the neighbourhood of a complex value which is not the root of a polynomial there is another value for which the absolute value of the polynomial is strictly less than the value of the polynomial for the initial complex number". It is, therefore, a simple formulation of the extreme value theorem for differentiable functions of a complex variable, an idea due to Legendre. But I do not intend to turn Argand into a prophet. By understanding the perfect match existing between complex numbers and polynomials, by building as Cédric just said a bridge, Argand did what was appropriate, which corresponds to what the Greeks called "suitability" or "symmetry"; in those times, it was in no way included in the "constructed" definition of a polynomial, but it has become natural.

Pierre Cartier: I would like to add a historical detail about complex numbers. In the 18th century, in an article on winds,[21] D'Alembert introduced complex numbers in a very explicit manner: he introduced two coordinates x and y in the plane, he did $x + iy$, and he expressed in fluid dynamics what later came to be known as Cauchy-Riemann equations.

Cédric Villani: He already saw the connection with complex analysis?

Pierre Cartier: Yes, in effect he uses a method which will be later taken up in aerodynamics in the 20th century. It uses (holomorphic) functions of a complex

---

[21] *Reflections on the General Cause of Winds*, 1745.

variable and thus performs a conformal transformation.[22] This enabled D'Alembert to study the distribution of winds on a large scale.

Jean Dhombres: In his study of winds, D'Alembert defined complex exponentials as though they were actions, and to explain this, I am going to use "I", thus reproducing a frequent habit among mathematicians (remember our school years: "I draw the line from A to B", "I set the operation..."). So a priori, I set $e^{ix} = a + ib$. I now make the two coefficients a and b depend on x by considering them as functions. I accept that the fundamental property of the exponential is to equal its derivative. I thus obtain $ie^{ix} dx = da + idb$, and as the left hand side is just $i(a + ib)dx$, separating the real term from the imaginary one, I get two equations from which it is easy to deduce that the function b/a is the solution of the differential equation of a tangent $(d(b/a) = (1 + (b/a)^2)dx)$. This introduces the angle and then gives Euler's formula $e^{ix} = \cos x + i \sin x$. It is a heavy-handed proof, which is so inductive that it does not seem to satisfy the axiomatic methodology. We agree to say it lacks rigor, when what is troublesome is not this but that the existence of the exponential as a complex number is presumed, and that this presumption leads to a unique object. The method, none other than that of Descartes and called the method of undetermined coefficients by D'Alembert, is probably the place where one best sees the mathematician's freedom of imagination, but also its limits, as though the real world were holding out against it. Something similar to the gesture of the fisherman in Indian thought who throws his net into the unknown, rather than logical thinking, led D'Alembert to set $e^{ix} = a + ib$ (who was not familiar with the then unpublished work of Euler). D'Alembert can be said to have brought back in his net the angle, a notion that is part of our everyday life, our experiments, our experience, albeit in a different guise, but not distorted, included in the construction of the field of complex numbers.

Pierre Cartier: Two small historical additions. We have talked of Euclid's rigor, and rightly so. But it is important to know that before the 17th century, and essentially before Euler and D'Alembert's work, the theory of angles did not exist. What is an angle of 1 °? To answer this question you need to know how to construct the regular polygon with 360 sides, which, according to Gauss, is impossible. Hence talking of an angle of 1° amounts to talking of something that does not exist. And in this context Euclid's first proof (the construction of the equilateral triangle) is wrong: I have a base, I draw a circular arc, another circular, they intersect, and I have won! But why do these arcs intersect? In Euclid's first proof, nothing in the axioms guarantees that the two circular arcs intersect. In fact, it is only at the of the 19th century that Pasch[23] and others made sure this works—which was later summarized in Hilbert's[24] geometry book on the foundations of geometry.

---

[22] Such as when drawing the map of some region of the Earth, angles are respected, but not lengths...

[23] Moritz Pasch (1843–1930), German mathematician, who wrote one of the first books axiomatizing geometry (1882).

[24] David Hilbert, *Foundations of Geometry* (1899), ed. Jacques Gabay, 1997.

Cédric Villani: Euclid supposed the two circular arcs intersect: it is intuitively obvious, but in fact it does not follow from the axioms of geometry… So how can the gap in the proof be filled? An analysis argument is needed, a study of the distance function…?

Pierre Cartier: Absolutely, one has to use analysis.

Cédric Villani: How sad…

Gerhard Heinzmann: The purity of methods has to be abandoned. Aristotle used to forbid going from "one genus to another" in a proof, "for instance, proving geometrical truths by arithmetic".[25] And though Descartes and others later applied algebraic methods to geometry, this ideal of purity remained largely untouched.[26] Among those who gave up the epistemological value of the purity of methods there is Henri Poincaré, for whom analytic methods are not only more concise but enable us to obtain new results in "geometry" (=topology) through association:

> "Indeed, the sole purpose of geometry is not to provide immediate descriptions of bodies falling within the scope of our senses: it is foremost an analytic study of a group; as a consequence, nothing prevents us to deal with analogous and more general groups. It will be said, why should we not preserve the analytic language and replace it by a geometric language, which loses all its benefits as soon as our senses can no longer be involved. The fact is this new language is concise; next that the similarity with ordinary geometry can create fertile associations of ideas and suggest useful generalizations".

Pierre Cartier: To found geometry, analysis cannot be dispensed with. Hilbert's book on the foundations of geometry—and even better, Artin's *Geometric Algebra*—does a superb job of showing that a whole geometry can be founded over an ordered field (algebraic numbers, rational numbers, etc.), but that to obtain unsophisticated, yet fundamental, results on angles, intersecting curves, etc., at some point, recourse to analysis cannot be avoided, in other words, the whole field of real numbers needs to be generated. And the latter cannot be defined without extensive recourse to infinity.

Jean Dhombres: In support of all that Pierre said about the unavoidable aspect of infinity when it is reworked by analysis, and this time going from geometry to algebra, I would like to return to Argand and his proof of the fundamental theorem. On the one hand, because he lacked a property ensuring that the extremum of the absolute value of a polynomial in a disk is effectively reached for some value of the variable (Cauchy too lacked this). Topology got really started in the last third of the 19th century when the lack of a correct proof for such a property was realized; on the other hand, because the manner Argand studied inequalities consisted in cuttings, now ordinary in analysis, by which we ensure that the omission of certain terms is

---

[25] *Posterior Analytics*, Book I, Part 7.

[26] Michael Detlefsen, Andrew Arana (2011), "Purity of Methods", *Philosophers imprint* 11 (2), 1–20, p. 5.

safe. In my view, the game of epsilons and deltas, which makes up the anthropology of the analyst, begins with Argand, even if it is with Weierstrass[27] that this became a general rule.

Gerhard Heinzmann: Historically, this change in attitude is remarkable. From the Euclidean model one passes to an analytic model and the tool changes the mathematical problems.

Pierre Cartier: This is what the whole changeover of the 19th century is about.

Cédric Villani: A changeover which is found in embryonic form in Euler's history: analysis is put back at the core of the entire structure.

The Nature of Mathematical Objects

Intuition: To Discover or to Invent?

Cédric Villani: In my view, the issue of the status of mathematics is not all that important. Is it Platonic (mathematical objects preexisting in a world of ideas), Aristotelian (mathematical objects used to describe the *real* world), etc.? What is important is not the tool, it is the way it is used... Of course, philosophical conceptions can influence the manner research is conducted in mathematics, in the sense that someone convinced there is something intrinsic waiting to be discovered will not pursue research in the same way as someone persuaded it is a manmade construction movement. Personally, I belong to the school of thought that posits a preexisting harmony and which, for each given problem, seeks for the gem, being convinced that it exists. I am one of the miracle seekers, but not one who creates it or seeks for something very clever in his own resources.

Gerhard Heinzmann: But what are the conditions for acceding to this truth with no clausal link with us? How can it be affirmed there exists something abstract? Either we allow ourselves an abstraction procedure, or we are forced to say "it is by a particular ability", some sort of unexplained intuition.

Cédric Villani: Yes, an unexplained intuition, a personal and almost religious conviction.

Gerhard Heinzmann: This is the weak point of this conception. The others, the constructivists, are more rational But perhaps they are less successful.

---

[27] Karl Weierstrass (1815–1897), German mathematician considered to be the founder of modern analysis.

Cédric Villani: Constructivists, who have less faith in miracles, may not rush in so many wrong directions, but this is to be seen... Anyhow, my way of proceeding, which always consists in searching for bridges between already existing things is more that of someone persuaded of a preexisting harmony.

Sylvestre Huet: Henri Poincaré said: "Logic ... is the instrument of demonstration; intuition is the instrument of invention".[28] Is this remark directly connected to your exchange?

Gerhard Heinzmann: Yes and no! Yes, because the order of invention may proceed without logic. Associations, intuition, imagination, dreams, fictitious entities: everything can be useful and justified as an invention tool. However, to justify, in other words to prove, for example the hypothesis that the study of some transformation groups leads to geometries with constant curvature, correct "logical" arguments are needed. Now, the difficulty resides in determining what is meant by "logically correct". Contrary to what the above quoted extract from Poincaré may lead to believe, for Poincaré logic was neither the only means to reach certainty, nor external to the realm of intuition. On the contrary, *pure* intuition also gives certainty and enables us not only to invent but also to *demonstrate*.[1] Poincaré was opposed to the logicians' thesis who claim to be able to demonstrate all mathematical truths without having recourse to intuition, once the principles of logic are admitted. Everything then depends on what is admitted as "logic". Since, with the extensive development of mathematical logic, the answer to this question is now more balanced than in Poincaré's days, Poincaré's interesting and modern argument against logicism is not so much the conjecture that there does not exist any purely logical transposition for each mathematical reasoning, but the affirmation that this transposition would lack epistemic values necessary to understand mathematical reasoning—and I think he is right!

Pierre Cartier: I largely subscribe to the thesis developed by Gerhard. I would happily use a metaphor, that of the motion of a pendulum. Watching it gives the impression of cogs and wheels fitted together in a complicated and precise manner, which enables it to start once the initial impulsion is given. This is equivalent to a logical system, and a very complicated computer program is an object of the same type. But the conception of the motion of a pendulum—or of a computer program, or even of the world's great clock according to Kepler and Newton—is the result of a different process, which cannot be reduced to a game of cogs and wheels. A long and complicated mathematical proof also requires preplanning, and hiding this from the reader may jeopardize his understanding. Is this not some sort of intuition? Does intention the same as intuition?

Jean Dhombres: At least in mathematics, constructivism flourished for a long time! This is what I was emphasizing by recalling the success, but also the weight, of

---

[28] Henri Poincaré, *The Value of Science*, p. 23.

Euclid's *Elements* as a structure. Can it be said that it is through intuition that Euler obtained his formula? We saw that the same formula follows from an impressive feat in D'Alembert! Anyhow, because of this formula, Viète's trigonometric formulas become trivial, or rather they can thereby be explained, and connections are forged between algebra and trigonometry coming from geometry. In return, is it not possible to think that Viète's formulas, which made him burst with joy, because of their very complexity have pushed mathematicians to find the hidden reasons? The motivation to simplify ties up with the idea that what is complicated is deficient. In my view, this motivation paradoxically justifies what Gerhard said about the role of logic. To take just one example: around 1807, after many pages of calculations, Fourier found what today is known as the Fourier series of a sawtooth function. The coefficients he obtains can be expressed simply, but their computation is awful. It is not logical. He, therefore, felt there must be a deep reason for this simplicity. As testified by his manuscript, he then immediately discovered the orthogonality relations: the whole theory could then develop analytically and he formulated in his "Analytic Theory of Heat". In order not to fail historical truth as well as not to prevent the understanding of what Karl Popper calls "the logic of discovery", he decided to include the computational stage.

Infinity: A Mathematical Necessity, a Physical Abnormality

Sylvestre Huet: Could you try to summarize this whole issue of the relationship between mathematical objects and reality, and why the ancient quarrel Plato/Aristotle is still going on? For example, Alain Connes made this unequivocal statement: "In my view, mathematics represents the only coherent strategy to understand and specify the external material reality unambiguously". In other words, he admits being at one extreme philosophically. In my professional experience, I noticed that this extreme position is shared by most "great" mathematicians. It is probably no coincidence. Anyhow, what is somewhat fascinating in the way in which you are presenting this relationship is, on the one hand, that notions of wonder and fascination stem both from discovering that a material reality can be mathematized in order to specify it better, possibly to understand it and *in fine*, to act; and on the other hand, the fact that self-contained developments become tools to specify this reality, as for example complex numbers that were in no way initially conceived to strike a significant chord with the real world.

In the evolution of mathematics and of their relationship with other empirical sciences, it is rather extraordinary that more and more complicated objects can be gradually mathematized, in particular coarse objects such as fractals used to describe the shape of clouds made of water vapor, snowflakes, or the coastline of Brittany... All this proves the awesome power of mathematics in the construction of other sciences (natural sciences, physical sciences, geography, etc.). How do you see this tremendous extension of mathematical tools? And, according to you, you who seem not very Platonic, is the ancient quarrel Plato/Arstotle still relevant?

Cédric Villani: In fact, I consider myself to be a Platonic mathematician, but I think this quarrel should not be continued, that it should be overcome. Mathematics should be addressed not in terms of conceptions, but in terms of approach, psychologically speaking. Everyone has different aptitudes, some are good at something, others at something else...

Gerhard Heinzmann: Platonism is a philosophical position which says that "mathematical entities exist autonomously, independently of the mathematician, and that the latter can reliably and indisputably access them". As I mentioned above, as a founding position, this position is controversial and, like all philosophical positions, has advantages and disadvantages. However, I get the impression that Cédric does not at all wish to defend Platonism as a philosophical position justifying his mathematical activities, but rather as a heuristic position to *do* mathematics. He is free to do so and a majority of mathematicians are probably Platonists in this sense, though they do not defend Platonism in a philosophical sense.

Sylvestre Huet: Let us take a concrete example... Pierre Cartier said we need infinity, which is a purely mathematical object, to construct basic mathematical tools such as real numbers. Now, the theoretical physicist's great fear is to find an infinity in his equations for he knows that at that "place", he will no longer be able to apply the laws of physics: in the real world, nothing is infinite. The obsession of the physicist will, therefore, be to get rid of it, even though he previously needed it in his arguments.

Cédric Villani: How can infinity be imagined? There are many ways of imagining infinity. In fact there are researchers who try to do without infinity—and it is far harder to work without infinity that with it. Infinity is an idea, an abstraction and a shortcut for a very large number, a number as large as wanted, which simplifies things. Historically, immense progress has been achieved by trying to do without infinity, by replacing it with numbers "as large as wanted" having quantitatively measurable, precise bounds; in contrast, immense progress have been achieved using infinity. Instead of being afraid, a physicist encountering a singularity (something discontinuous, a punctual change of the model's properties, somewhere density becomes infinite... a black hole, the big bang, or anything else) thinks: "This is where the thing is hiding". For example, Landau[29] discovered "Landau damping" by studying the singularities of the Vlasov equation in plasma physics, etc. I am sure that it is possible to find numerous examples. Thus, for those with a certain mind set, who are very pragmatic and want to build tools, doing without infinity becomes a major challenge; and for others who are seeking the key to the mystery, the idea that there are phenomena in the real world governed by some sort of "infinite idea" supposedly existing in a certain universe is also very important. In short, I think that it is necessary to move beyond the discussion "is mathematics like this or like that"?

---

[29]Lex Davidovitch Landau (1908–1968), Russian theoretical physicist, who received the Nobel prize in physics in 1962 "for his pioneering theories on the condensed state of matter, in particular liquid helium".

Sylvestre Huet: Are we to understand that the Platonic/Aristotelian position depends on the way mathematicians work? In other words, when Alain Connes says "I explore an external world…", the important word is not his Platonic standpoint about the fact that there are supposedly mathematical objects existing independently of man or of material reality, but the verb "explore" which describes his way of working.

Cédric Villani: Absolutely.

Pierre Cartier: Besides, this can be illustrated using fractals. Even if I have some reservations about Mandelbrot,[30] his stroke of genius should be recognized: He *saw* what was before everyone's eyes and what no one had seen before him. Yes, these wretched fractals were there in front of us! They are everywhere, in painting, in clouds represented on the photo hanging on this wall…

Now, mathematics can sometimes give rise to a certain myopia if one is focussed on something too precise, forgetting what is beside it. It is well-known that great discoveries—not only in mathematics—are often things that are looking us right in the eye, but that nobody sees, and all of a sudden: the king is naked! What was visible to everyone is discovered and explained to others.

Cédric Villani: It is like this idea which lasted for a long time that all functions are differentiable; even Cauchy at some point though that continuous functions were always differentiable…

Pierre Cartier: Galois also used to believe this.

Cédric Villani: Galois too? I did not know. And then one day non-differentiable functions were constructed. Initially, everyone said: "What is this, it is not possible?", then finally everyone changed their mind: "But yes, not all functions are differentiable, we have been using them for centuries and we never realized!".

Pierre Cartier: This hesitation is understandable because we are always a bit frightened in front of infinity. Now, when you want to construct a non-differentiable function or fractals, you have to iterate infinitely many times. If you stop halfway, you're done for. On the other hand, I gave a student from Strasbourg a master's thesis subject based on Linnik's[31] and Erdös'[32] work on stochastic properties of the distribution of prime numbers: I asked him to simulate Brownian motion using prime numbers. The student showed me his result, a curve, and I immediately replied, shocked: "No, it is not a Brownian motion". I could not readily explain why—it greatly swerved

---

[30]Benoît Mandelbrot (1924–2010), French-American mathematician, famous for having developed a new class of mathematical objects: fractal objects (or fractals).

[31]Yuri Linnik (1915–1972), Russian mathematician whose work was mainly in number theory, probability theory and statistics.

[32]Paul Erdös (1913–1996), Hungarian mathematician who largely contributed to the development of number theory and combinatorics.

and then returned too slowly. We then analyzed his result and understood where the mistake came from because we are rational beings: he had simply wrongly copied a formula. With a minor correction, it worked. I want to show through this example that it is possible to get the intuition of what a Brownian motion is, of what a fractal is. In the case at hand, my intuition resulted in a shock: the student's result did not correspond to what I was expecting. This is both one of the difficulties and one of the beauties of mathematics: to know whether there are mathematical ideas in some empyrean or other, but anyway there is a certain mathematical reality with different realizations—either in the calculations performed, or in the images under study and which strike you, raise questions, and force you to answer them. Still there is something in front of us. Even if I do not believe in the ideas of the cave which would have us restricting ourselves to observing the shadows of reality of another world beyond us, I am sure that a mathematician does not act in a void. Though I am not a formalist—anyhow not too much of a formalist, just what is needed...–, I recognize the power of formalist reasoning, of the formalization of language and of its extensive logical contributions in the 20th century. But formalism does not exhaust everything because on occasion there is nevertheless this element of surprise or shock which expresses something deep, and which is not easy to analyze. A rational analysis is always possible after the fact, but it does not tell the whole story.

Cédric Villani: This is why Mandelbrot is nonetheless an important figure: he put back intuition at the heart of the debate, its sensual and visual aspect, at a time when this was badly regarded.

Pierre Cartier: I spoke just after Mandelbrot at the yearly meeting of the (French) Association of Mathematics Professors in Public Education, and in fact, I was surprised that we both said the same thing: "Do not be afraid, use drawings". It was quite daring to say this in the early 1980s in front of 2000 professors, in the temple of the Association, at a time when using drawings was not rather unacceptable!

Gerhard Heinzmann: But obviously, intuition should not be only seen in drawings, that is to say in its sensitive form. It is also present in a non-sensitive form in the understanding of a proof. It is legitimate to ask ourselves whether all can be formalized? And, indeed, the understanding of a proof cannot be formalized, under penalty of a regression to infinity. It requires some sort of intuition, understood to be the overall view of a "know how". Of course, it a step-by-step proof is possible, but this does not amount to an understanding of a proof.

Cédric Villani: In a proof, we usually mention: "This is important, this and this". We don't only give the logical design which solves the problem. For the listener to understand, we extract the key points that will enable him to convince himself of the truth of the proof and to recreate it, when required. This separation between what is "important" and what is "secondary" is not logical (for logic dictates that all is important!), but intuitive, calling upon usual diagrams, reflexes, analogies, summaries...

Gerhard Heinzmann: Precisely, this understanding forms a broad field in the hermeneutics and analytics of philosophical approaches.

Jean Dhombres: As is very often the case, history teaches us modesty in our epistemological conclusions, by not taking sides, leaving all their importance to the answers that have been brought above. The first time infinity played a role as such in a mathematical proof is in a 1795 paper of Laplace. By skillfully introducing a parameter that Gauss will take up again in 1815—forgetting to say where he had found it!—, Laplace proved the existence of a function from the reals to a finite set (in fact pairs of roots of a given polynomial equation). He did not at all know how to compute this function, and could not in any way draw it, and equally had no intuition about this function: yet the infinite characteristic of real numbers asserts that the said function will take the same value for at least two distinct real values. This is sufficient for Laplace to finish his proof of the fundamental theorem of algebra, without having any means to effectively construct roots.

Aesthetic Criteria in Mathematics

Cédric Villani: The issue of aesthetics in mathematics, which in fact was fundamental in the exhibition *Mathematics, a sudden change of scenery*,[33] is equally essential. Visitors to the exhibition were struck, but also very surprised by the beauty of the exhibits. The general public indeed finds mentions of the aesthetic aspect of mathematics surprising, whereas for mathematicians, this is trivial…—they are even surprised that they are asked about it!

Jean Dhombres: An elegant proof or a beautiful proof…, these words have meaning in mathematics, but which practitioners keep deep inside them.

Gerhard Heinzmann: It is nevertheless necessary to explain why a proof is beautiful. On this point, there is something in Poincaré that sets him apart from tradition: he regarded as aesthetic what has greater possibilities of applications, in other words of becoming possible models:

> "the mathematician must have worked as artist.
> What we ask of him is to help us to see […]. Now, he sees best who stands highest".[34]

> "For a construction to …serve as stepping-stone to one wishing to mount, it must first of all possess a sort of unity enabling us to see in it something besides the juxtaposition of its elements. …A construction, therefore, becomes interesting only when it can be ranged beside other analogous constructions, forming species of the same genus".[35]

---

[33]Exhibition *Mathematics, a sudden change of scenery* (Cartier Foundation) welcomed 80 000 visitors between October 21, 2011 and March 18, 2012.

[34]*The Value of Science*, pp. 77–78.

[35]*Science and Hypothesis*, p. 17.

In his opinion, topology (*Analysis Situs*) is the most aesthetic subject because it has applications everywhere:

> "[in] the study of curves defined by differential equations and to generalize them to higher order differential equations and in particular in the study of the three-body problem. [He] needed it to study non-uniform functions of two variables, to study periods of multiple integrals and to apply this study to the development of the perturbation function. In short, [for him], *Analysis Situs* was a way to approach an important problem in group theory, namely the search for discrete groups or finite groups contained in a given continuous group".[36]

This vision of aesthetics has a modern sense which stands out from the traditional science of beauty or from the criticism of taste; it includes a cognitive function: a proof is all the more aesthetic when it can be used in other fields.

Jean Dhombres: It is a beautiful answer in tune with its times, a time when philosophical utilitarianism had the upper hand; one and half century earlier, Newton would have probably mentioned the aesthetics involved in the practice of a unique method leading to the unification, if not of science, at least of one of its sectors, for what will become celestial mechanics. As for the eccentric English mathematician Godfrey H. Hardy, he mentioned the aesthetics of all mathematical theories having no applications! And for André Weil, aesthetics resided in a kind of metaphysical shiver that takes hold of the inventor, in a hurry to get rid of this beautiful weight by formalizing it for common consumption.

Pierre Cartier: The Hungarian mathematician, Paul Erdös,[37] a highly colorful character, used to talk of the proofs "in the Book". He meant those that have such a resonance, or such an element of surprise that nothing can surpass them and that they need to be included in an anthology [such an anthology exists, published by Springer]. Such a proof needs to go right to the point, while revealing some unexpected aspects. The law of quadratic reciprocity provides an example; Gauss gave six different proofs for it; the simple character of each is stunning, and so is the diversity: a book of hours!

Grothendieck has a somewhat different idea: to develop the natural foundations in such a way that difficulties dissolve, like the sea flooding the beach and becoming "etale" (one of his fetish words). In this sense, the most beautiful proof is the one about which one can simply say: "See!", which harmoniously fits in the structure. Grothendieck started his scientific work with "Functional Analysis", a field where Gerfand[38] had showed the way. The discovery after 1945 of Gelfand's thesis on "normed rings" gave rise to a great sense of wonder. It contained difficult and deep theorems—N. Wiener's Tauberian theorem, G. Birkoff's ergodic theorem—like so

---

[36] See Henri Poincaré, Analyse des travaux scientifiques de Henri Poincaré faite par lui-même, *Acta Mathematica*, 38, 1921, pp. 1–135, p. 101.

[37] See his biography in *Erdös, the man who loved only numbers*, Berlin, 2000.

[38] Israel Gelfand (1913–2009), Russian mathematician, student of Andrei Kolmogorov, whose work is related to every mathematical field. He was, however, mostly known for his contributions to functional analysis and their impact in quantum mechanics.

many ripe fruits of a new theory harmoniously unraveling while retaining a high degree of abstraction: the eagle that descends from the sky upon its prey.

Aestheticism has pitfalls. In mathematics, a proof that is too beautiful is not often very *pliable*; it does not lend itself to modifications, it can close a path.

Sylvestre Huet: And is this aestheticism the same for all mathematicians?

Pierre Cartier: Mostly.

Gerhard Heinzmann: It is not about value, but about cognitive quality. A proof using many *ad hoc* elements and only usable in the field considered is less aesthetic than a proof applicable to several fields, for example like proofs in model theory.

Cédric Villani: Poincaré talked about the value of aestheticism as something that enables it to become useful, harmonious, but also as something the enables to guide mathematicians. These feelings are widely shared.

Sylvestre Huet: Are there other approaches of the aesthetic aspect of mathematics apart from that of Poincaré? In particular, it is not unusual to hear that a proof is elegant because it is short, clear, and uses very few axioms... Are there rules that need to be upheld, criteria like Ockham's razor?

Gerhard Heinzmann: The symptoms you mention must indeed be present to be able to talk of a mathematical aestheticism, independently of its theoretical conception. In this regard, in her book "Aestheticism and Mathematics",[39] Caroline Julien distinguishes between the Poincaré solution and the Platonic solution: according to the latter, the idea of beauty merges with the truth of mathematical facts that exist independently of the mind. On the contrary, for Poincaré, "beauty is not an idea in the Platonic sense, but a category that enables us to explain the pleasure of the senses". Nevertheless, American philosopher Nelson Goodman (1906–1998) is the one who profoundly influenced the history of aestheticism in the 20th century by replacing the question "what is art" by the question "when is art?". His answer thus contains technical and precise symptoms which answer your question: "simplicity", "saturation", and "exemplification" are symptoms (too technical to be clarified here) interpreted in the case of mathematics by Caroline Julien in her book.

Jean Dhombres: The "aestheticism of simplicity" may be classical, but, in mathematics, one also finds an aestheticism that could be qualified as baroque. The simplest example I can think of is found in John von Neumann's approach, which introduced the notion of a Hilbert space, concerning the mathematical foundations of quantum mechanics. Two models of quantum mechanics were available to him in his days, both were satisfactory realizations of experimental data. One of them (that of

---

[39]Caroline Julien, *Esthétique et mathématiques. Une exploration goodmanienne*, Presses universitaires de Rennes, 2008.

Heisenberg) used matrices, and was, therefore, a matter of the discrete character of numerical algorithms. The other one (that of Schrodinger) used wave functions, and fell under continuity and differential analysis. Von Neumann's viewpoint, which I would call aesthetic and baroque because he brings order into objective disorder, was to say that these two theories, of very different type, were both a matter of the same structure... which still needed to be found by suitable selection. And this is where construction intervenes. In this matter, what is it that is said to be beautiful: the idea of a common structure, or the determination of this structure?

Determinism, Chaos, and Predictability

Sylvestre Huet: How do you feel about the current state of discussions started 30 years ago on determinism and indeterminism, non-linear dynamic systems and deterministic chaos? Has this approach been fruitful for the sciences in general and mathematics in particular?

Pierre Cartier: There is no doubt that for celestial mechanics this was a major step forward, confirming Poincaré's intuition and paving the way for a new methodology to study the long-term evolution and stability of the Solar System. Many conventional ideas were knocked down. For the rest of mathematics, there is a small domain named "Study of dynamical systems" where ideas of this type are developed, but I do not think it has really irrigated the entire field of mathematics—contrary to that of physics.

Generally speaking, the discussion on determinism and indeterminism is multi-faceted and difficult. In Lapalace's perspective, an "omniscient" mind could clearly foresee the future—and recover the past. More realistically, information comes at a cost; the economic assessment has to be borne in mind: cost of the information to be collected versus benefits of forecasts made by means of this information.

A particularly striking example was discovered by J. Moser in the 1970s. It relates to an astronomical system consisting of two massive stars, rotating around a common center in accordance with Kepler's laws. A low-mass comet is traveling along the perpendicular axis to the plane of the two orbits, passing through the center of the system. Let us take the revolution period of the two stars as our year. Question: for one thousand consecutive years, it has been observed that the comet passes through the center. Will it come back or not the following year? Answer: no answer can be given, the reason being that the answer may depend on the 50,000th decimal place of the number giving the initial velocity of the comet, and that computing this number to this decimal place would require incredibly long observations. I don't have enough space here to develop all new viewpoints about Darwin's theory of evolution, about the foundations of thermodynamics and of the theory of gases,[40] about teleology, and finally about the supposed indeterminism of quantum mechanics. It is a very open

---

[40]Maxwell imagined a demon opening a door for fast molecules and closing it for slow ones. Precise experiments simulate such a demon.

field, and the debate between chance and necessity has become far more positive and fruitful.

Cédric Villani: Nowadays these concepts have become part of our common culture. They are the means of interpretation of numerous phenomena... Nonetheless, as Poincaré and then Lorenz fully realized, this theory is based on two pillars: the first pillar, exponentially growing uncertainty which makes all precise long-term forecasts impossible, is remembered, but the second one is, on the contrary, often forgotten: statistical predictability, enabling us to assert that this or that will happen with a certain chance of success... The greater the chaos, the better statistical predictability.

Jean Dhombres: I would like to add an anecdote, just to show that mathematicians are truly at the origin of the debate "determinism and indeterminism" when probability was seriously implemented in natural philosophy. In the 1758 second edition of his *Traité de dynamique* D'Alembert gave "proofs" for the principle of inertia, which is a form of simplified and absolute determinism. Eleven years later, the very young Laplace, a student with a barely one-month old masters degree from the University of Caen came to Paris; as he failed to be received by the academician, he wrote him a letter to tell him he had made a double mistake: the determinism of the principle of inertia cannot be proved by mathematics, nor by metaphysics.

But determinism had to be set out as a condition in order to do science, in particular to be able to turn probability into a tool to search for the causes of certain phenomena. His expressly unsubmissive manner won him D'alembert's trust, who found a professorship for Laplace in the *École militaire* of Paris... but did not change his dogmatic standpoint. Laplace never mentioned it again, but gave determinism his most famous statement, while insisting on the "delicate" handling of probability. It was no longer a question of "given that" which may characterize the Euclidean world, but of "do as if".

Sylvestre Huet: Besides this debate took on a very societal dimension, especially fed by a total distortion of Isabelle Stengers'[41] and Ilya Prigogine's[42] message. Actually the latter proved that Laplacian determinism—the demon that would be capable of forecasting the evolution of any system by knowing its laws and initial positions— cannot exist in a great many dynamical systems for they are not linear. Moreover, The French astronomer, Jacques laskars' work showed that this extends to the celestial mechanics of the Solar System, and that it is not possible to compute the future precise position of the Earth beyond some hundreds of millions of years. Nevertheless, for both researchers, this in no way amounts to an abdication of the human mind, on the contrary it represents a progress since it is now possible to compute forecast

---

[41] Isabelle Stengers, born in 1949, Belgian philosopher and historian of science.

[42] Ilya Prigogine, born in 1949, Belgian philosopher and historian of science of Russian origin, who received the Nobel Prize in chemistry in 1977 for his contribution to the thermodynamics of irreversible processes and to the theory of dissipative structures. In particular, he showed that when matter moves away from its equilibrium state, self-organization occurs—a phenomenon that can be observed in physics, biology and climatology.

horizons of systems whereas before the illusion of infinite predictability prevailed. This thought has been distorted and is often presented to the public in a caricatural form: "Laplacian determinism is a mistake, hence nothing can be predicted, and if nothing can be predicted, it is not possible to set out a political program for the transformation of society".

Jean Dhombres: This was also Bruno Latour's Nietzschean position, that of absolute relativism, all knowledge being merely a pose, for it hides the nature of social forces and the interests of all sides that it may impose it. But the tide has been turning, and the Sokal case, as well as Jean Bricmont's comments, have had an impact; he has come back to feelings said to be of empathy with scientific practice.

Cédric Villani: It's a dreadful discourse, which is still very much part of social sciences —and also other fields—, and which consists in saying: "Since it is very complex and we understand nothing, we should do nothing, or else we should change everything". But when you ask those who say that "we should change everything" what it is that it should be replaced by, the answer is never very clear.

Jean Dhombres: One could also turn the argument around: since you got it wrong once—and one always gets in wrong once—, you are always going to get it wrong. This is a perfect sophist's argument; and it is hard to counter without sinking into complacency, which is a bit the hallmark of scientism. This complacency sometimes lies in wait for mathematics, a science that has historically recognized insufficiencies, but never mistakes which would amount to contradictions and would have been accepted by a generation of mathematicians. It, therefore, also lies in wait for logic since the latter is largely mathematized. It is actually in the name of this sophism that Galileo ended up being condemned: by pulling down even a part of the Aristotelian construction, Galileo was seemingly undermining all truths, and thus creating chaos.

Gerhard Heinzmann: Indeed, going from sophism to scientism would be as absurd as giving up perception because of illusions, paradoxes or physiological anomalies, or as considering intuition inadequate because of its fallibility.

Cédric Villani: Ultimately, this is an issue in the moral sphere. A politician may make mistakes, but it is his duty to elaborate a project, it is his mandate. He has a vision, he tries, he adapts, but if for any reason he resigns then he no longer does his job. This goes beyond the issue of science.

Political and Social History of Mathematics Education

The Role of Mathematics in the Education System

Sylvestre Huet: We are now going to explore three major points concerning aspects of "mathematics and society": education and training, mathematics and industry, and finally the use of mathematics in political debates and generally speaking in the humanities.

I suggest we start with the issue of training. In France discussions on education and mathematics raise passions—of even epic dimension—and show how sensitive the issue is. The particular status of mathematics is related to the characteristics of our education system: dichotomy between general education and technical education, and hierarchy between the different types of general education. Mathematics—and especially the marks obtained in this subject—play an absolutely key role in our education's selective system. Much has been written about this and will continue to be written. What are your views about this issue?

Pierre Cartier: In this chapter, I am going assume the role of grandfather... I experienced the first phase of the transformation in the 1950s, because I was a preparatory class[1] student at the *Lycée Saint-Louis* (in Paris) in 1949, withe good schoolmasters teaching us "old" mathematics. I remember one of my mater sniggering one day when he saw me reading a book on topology—it was one of the first Bourbaki books.[2] At the time, the problem of education was that for historical reasons, probably difficult to analyze, French scientific education in the period 1930–1955 was lagging consid-

---

[1]Two year preparation for the entrance examination or rather *concours* for the French grandes écoles. Translator's note.

[2]Nicolas Bourbaki, pseudonyme given to a group of eminent French mathematicians who, starting in the 1940s, undertook the publication of an extensive treatise called "Elements of mathematics". Its most famous members include André Weil, Jean Dieudonné, Jean-Pierre Serre, Alain Connes, Pierre Cartier...

© Springer India 2016
P. Cartier et al., *Freedom in Mathematics*,
DOI 10.1007/978-81-322-2788-5_3

erably behind. When I obtained the Advanced Teaching Degree in mathematics,[3] in 1953. Henri Cartan[4] used to make fun of any teaching that said that matrices would not be taught because matrix product was not well-defined...

The same held in theoretical physics: let us not forget that the first coherent course on quantum mechanics was given by Albert Messiah in 1963–1964 in an almost clandestine manner (a bus used to pick up students *place de la Sorbonne* to take them to Saclay). When I followed the course in general physics at the start of the academic year 1950 at the Sorbonne, the first class began with: "Gentlemen (no notice was taken of the handful of girls attending the course), there is no room in my class for what some call the atomic hypothesis". Five years after the Hiroshima atomic bomb! Let me tell you that I never set foot again in this class! I went to see Pierre Aigrain, who was my mentor at the *École Normale*. He laughed and said: "You still need to get your certificate at the end of the year, so make some minimal compromise and I will teach you modern physics". The big handbook of the time, the "Bruhat"—which had many other qualities—mentioned blackbody radiation and Planck's law in very small print in the fourth edition.

Sylvestre Huet: Can you give more details about the nature of this backwardness in scientific education? It lagged behind with respect to whom, to what?

Pierre Cartier: With respect to scientific developments! Secondary and university education took no notice of all that had happened in physics (with Einstein, Planck...) and in mathematics (Hilbert, Weyl...) in the early 20th century. There was only one way out: to turn everything upside down. Regarding the outcome, let us take the case of "modern" mathematics, with widespread abuses that I was the first to acknowledge. [Having had a short experience as a secondary school teacher for a year, after passing the Advanced Teaching Degree, I think I was realistic than others about this debate— and then I came from a family of teachers.] All of a sudden, all needed to be changed. And as usual, the baby was thrown out with the bathwater. The pendulum always swings from one extreme to the other...

Cédric Villani: To comment on what Pierre just said, if we look at the current situation and ask ourselves whether there is any mathematical specificity, it can be said that, undeniably there is a specificity; it consists in the period the mathematics taught is from. Indeed I think that it is the only high school subject where one is so far removed in time from the object being studied. Let me explain: in history, modern history is taught, in physics and chemistry, 20th century concepts are taught, in geology, continental drift, a mid-20th century discovery, is taught, in biology one learns about the DNA which is also a recent concept... In mathematics, all that is taught in high school dates from before 1800 on the whole. Modern mathematics were

---

[3] Known as *Agrégation*, a degree specific to France, needed to teach in *classes práparatoires* or in a good high school. Translator's note.

[4] Henri Cartan (1904–2008), French mathematician, founding member of the Bourbaki group, considered to be one of the most influential French mathematicians of his times.

probably an attempt to overcome this lag, at least formally, to use a modern form for some fashionable present-day thing. One could obviously argue: does mathematics lend itself to it or not, is it intrinsic to the subject, is it necessary to study something from so long ago? These are tricky questions... For example, I am among those who, for that matter like Ngô Bao Châo,[5] publicly expressed their attachment to Euclidean geometry. As a kid, I used to find the geometry of the triangle truly fascinating, and I continue to think that it is the best training for logical proofs. It is something very old, which is useless, in the sense that it does not give rise to any activity—even for a professional mathematician. It is albeit remarkable as a school of thought. This specificity is part of mathematics, at least in my conception of it, where the foremost purpose—I did not say the only one—of a mathematics course is to develop the method rather than the object. The object is irrelevant, learning to reason, to give an imaginative, rigorous reasoning, to persevere on a problem with the aim of finding how to access it, etc. Of course, this cannot be the only purpose, and indeed, the consequence of teaching old things is that people coming out of high school are not even aware that mathematics is connected to daily life, to motion, transport, innovation, technology. They think that it is some sort of language, a bit like an old language, like Latin, where the main purpose of the object is not defined, and a cultural heritage in some ways.

And then, aspects—already mentioned—related to the dialectic with the real world are not always understood, the fact that there is an attempt at formalization, that abstract concepts and mathematics often stem from concrete phenomena, all this may seem dry and unmotivated.

Therefore, all these aspects have to be taken into account when talking of mathematics education. It is not necessarily easy, the correct balance needs to be found, seemingly contradictory aims have to be reconciled, that it is necessary to learn to reason on simple matters, yet awaken the awareness of future citizens about the fact that one cannot do away with mathematics in research, in technological as well as social advances.

Jean Dhombres: We could present a brief history of mathematics education in France. In fact, under the monarchy (16–17th century), no one with the title and career of a mathematics teacher can be found. There are no "M. Dupont, mathematics teacher" before the French revolution!

Pierre Cartier: Even in military schools?

Jean Dhombres: Yes, those who do mathematics in such schools fall under mechanics, fortifications, what could be called "applied mathematics". Only the entrance examination to military schools are about mathematics, and as a result those who write textbooks, Camus, Bézout, or Bossut, all academicians, control the program which defines an ability, not a culture. Some of these men had the idea of including mathematics as a compulsory subject in the revolutionary program—from October

---

[5]Ngô Bao Châo, Franco-Vietnamese mathematician, jointly awarded the Fields Medal in 2010.

1789. Thus France is the first country where mathematics became a compulsory subject alongside with the humanities for those following a secondary education—they were not many. The first mathematics course is given in the *École Normale* in the Year III[6] for future teachers, who will teach Greek as well as mathematics. The 1802 law (article 1) for the establishment of high schools stipulates that "Latin and mathematics will taught in high schools". Besides, the revolutionary ideology concerning mathematics also has to be taken into account, in other words the idea that, under the monarchy, mathematical forms of thinking were the least marred by class relations, and the farthest removed from religion. It is no accident that Lazare Carnot became Director (under the *Directoire*[7]), then Minister, that Gaspard Monge was Minister of the Navy during the first republican government in 1792, that Pierre-Simon de Laplace became Home Secretary in 1799...

Cédric Villani: At the same time, Napoleon is a member of the *Institut de France*.

Jean Dhombres: It is interesting that part of this ideology—which takes us to modern mathematics—considered that the "good" mathematics that should be taught is the mathematics connected to ongoing research, in other words mathematics that was then being done. The reason for teaching Monge's work, namely descriptive geometry, is mostly because it was a new field of research and because textbooks of the time explicitly mention that new mathematics have to be taught.

Cédric Villani: Yes, it is incredible, Monge said: "There is a need for a class on descriptive geometry at the *École Normale Supérieure*". I read that out one day to flabbergasted student of the ENS—the prospect of a class on descriptive geometry must have seemed terrifying to them...

Pierre Cartier: And yet, it remained part of examinations until 1960...

Jean Dhombres: The political will—which is not insignificant in the reform of modern mathematics—was to adapt French education to industrial society. The problem is that the ideology remained stuck in innovative mathematics... of 1800! Though some people reacted before 1960, in particular Henri Poincaré, there was some sort of immobilism. The backwardness mentioned earlier by Pierre can be defined with respect to what had been the leading trend in mathematics education in France. Finally, it is a matter of backwardness with respect to ourselves.

Pierre Cartier: To move away from this immobilism, a strong political will aimed at a complete overhaul of university teaching was needed. André Lichnerowicz,[8]

---

[6]Of the revolutionary calendar. Translator's note.

[7]Form of government of the First French Republic, which lasted from October 1795 to November 1799. Translator's note.

[8]André Lichnerowicz (1915–1998), French mathematician, Professor at the *Collège de France* and member of the French Academy of Sciences, in particular known for his chairmanship of the

author of innovative undergraduate textbooks, a specialist of Einstein's relativity theory, was the linchpin of the famous Caen Colloquium (1956) which attempted to define new foundations for university training.

Jean Dhombres: Precisely. And it then emerges that Lichnerowicz had a very distinctive ideology, which is not that of 1800. There was a change.

Pierre Cartier: First of all, Lichnerowicz was a rightist, from the catholic right. He was a surprising example of a "conservative reformer"—best exemplified by Valéry Giscard D'Estaing.[9]

Jean Dhombres: He also assiduously attended the meetings of Catholic thinkers at the *Mutualité*,[10] and as such represented a new breed in the French scientific world, somewhat shriveled up since the old fights of 1900 for secularism. The idea underlying all this was that a radical reform of mathematics education would prevent the transmission of cultural assets within families. In this way, the sons of engineers, of the upper middle-classes, steeped in a certain culture, would have been on an equal footing with others. Everyone on an equal footing! This is the ideology behind this reform. One wonders why it only concerned mathematics, when it applies likewise to biology, physics, chemistry education.

Cédric Villani: On this point, I would like to throw around another question which in my view is equally characteristic of the reality on the ground in mathematics. It is the comparative disconnection of mathematics teachers from other teachers. They often find it difficult to come together on common projects, or of forging a link with other subjects for their students. It is important to recall that the mathematical sciences (I insist on the word "sciences") often give remarkable results through interaction with the other sciences, that some of their most impressive applications and most profound motivations can be found in the latter, both for researchers and students.

Gerhard Heinzmann: What you have just said is very interesting. I come from a different model—the German model—and have always wanted to study mathematics, but have not come across the same selection criteria. On the contrary, I was told: "You will first study Greek and Latin", like in France in 1900. Hence I went to a "humanities" high school as they are called in Germany. At university, I was enroled in Albrecht Dold's topology class, a monument of abstraction, and I noticed that my classmates from mathematical high schools had greater difficulties understanding than I had because in some way they were "distorted" or "formatted" to a style of

---

(Footnote 8 continued)

mathematics education reform commission, the so-called *Lichnerowicz Commission*, from 1967 to 1972, consisting of the teaching of formal (so-called "modern") mathematics with an axiomatic base (algebraic structures, vector spaces and set theory) as early as primary school—without any success.

[9]French president from 1974 to 1981. Translator's note.

[10]A multi-purpose hall in Paris. Translator's note.

reasoning. But, contrary to France, in Germany, it does not matter if you are not well versed in mathematics. There is no such system of competitive entrance examinations with a selection based on mathematics. In France, before the *École normale supérieure*, the *École polytechnique* was the main pathway for all mathematicians—incidentally, Germans have much envied France for this mathematization of the selection system, inherited from Auguste Comte's positivism. In Germany, it was rather the Kantian tradition of idealism which predominated, philosophers used to teach mathematics—Kant taught mathematics—and mathematical studies used to be conducted in the faculty of philosophy.

Generally speaking, mathematics education in French schools carries undoubtedly more weight that in Germany and monodisciplinary studies in France lead to greater specialization. Take me for example: I held a scholarship from one of the three main German foundations (Cusanuswerk), I came to study in France in 1973/1974 after obtaining my Bachelor's degree in mathematics and philosophy, thanks to a grant from the French government, but I suffered enormously, compared to my friends from the ENS with whom I followed classes in mathematics in Paris VII. ENS students are far better trained than student from German universities. In France, the mathematical model in competitive examinations is obviously very important, because the entire elite goes through these unified competitive entrance examinations (*École normale supérieure, École polytechnique, École centrale*, advanced business schools, etc.).

From the Abstract to the Concrete?

Cédric Villani: I would like to follow up on what Gerhard said about German methods. The way he put things, interesting questions arise about the relationship with abstract thinking. In the French tradition of presentation, usually courses go from the abstract to the concrete. But in general this is not historically correct; this is not how things acquire meaning in people's minds, and on occasion it is good to begin with something concrete, and then to draw out abstractions. And this is sometimes more efficient for a comprehensive understanding.

On the issue of the relationship of mathematics and the concrete world, and of the difference between the German and French standpoints, during the exhibition *Mathematics, a sudden change of scenery*, I remember being struck by French journalists going into raptures over a series of short films juxtaposing mathematical formulas and illustrations of concrete phenomena—a leopard spot and a reaction-diffusion equation, a flame and a heat equation, etc. They were saying: "Ah, its incredible, so what one learns in mathematics has really to do with these phenomena!" What the German journalist present felt was precisely the opposite; he was saying: "But this never ending story is so commonplace, mathematics and the real world, everyone knows that!" And I asked myself whether a cultural question lies behind this, a way of looking at mathematics education and pedagogy. Perhaps Gerhard could comment?

Gerhard Heinzmann: Indeed, in Germany, David Hilbert was a turning point that led to the end of the separation between pure and applied mathematics and of the illusion that mathematics represents Platonic ideas. From a Hilbertian standpoint, applied mathematics differ from pure mathematics only because they provide an interpretation for axiomatic schemes (formal formulas) which replace traditional axioms in pure mathematics. But here too, I am talking as a logician and avoiding the substance of what Cédric asked, namely whether the cultural environment influences mathematics. Though this is certainly the case, this proposition is useful only if the exact links between mathematical theory and cultural practice can be specified. In the case at hand, this link is all the more difficult to determine for since the 1930s, mathematics in France have been dominated by the Bourbaki group, itself strongly influenced by Hilbert. Hence one should rather analyze how the image of mathematics is transmitted in both societies. On the German side, I am thinking of the writers Hermann Broch, Robert Musil, Hermann Hesse or Max Frisch in whose work mathematics plays a central role. On the French side, I think of Paul Valéry who was an admirer of Poincaré and of the cubist and abstract painting movements (for example François Morellet).

A key to generally interpret a social difference influencing the image of mathematics is, on the one hand, the fact that in Germany mathematics is generally studied together with physics as a secondary field, and on the other hand, that research and teaching are or have been far more interconnected than in France where competitive entrance examinations determine programs in comparative independence from research.

Cédric Villani: Journalists always ask me two questions: (1) what are mathematics useful for? (with obviously the underlying idea that they are useless!); and (2) how did you develop a taste for mathematics? (with the idea that it is anyway incredible that it may be found fascinating, or just simply interesting!). These two questions are of course connected.

Gerhard Heinzmann: We have answered this question in the first part of this book: it is the only concrete language, or rather exact language, at our disposal to formulate our theories about the world.

Cédric Villani: Abstract, but yet so concrete finally!

Gerhard Heinzmann: Indeed, even DNA folded one thousand times can be observed under the microscope. But if topological methods are used to describe the double helix of DNA, folded back on itself in the nucleus of a cell inside a chromosome, everything becomes clear... with nevertheless the possibility of making a mistake: indeed topological methods may not provide an adequate "model" for that biological fact. An observation or another theoretical fact contradicting the result obtained by topological methods would call everything into question.

Mathematics: A Tool for Democratization or for Social Reproduction?

Jean Dhombres: We have touched upon an essential aspect of the issue of mathematics education, namely the revolutionary idea of a meritocracy. The emergence of a meritocracy is concomitant with the French revolution, and France was the first country to create a solely mathematical selection mechanism to select its technical elite—*Polytechnicians*, the *Corps des Ponts et Chaussées*,[11] mining engineers. The only question asked in the oral examination was a mathematical one. During the two revolutionary years, Year III and Year IV, there used to be a paper on morality, which merely consisted in obtaining an attestation from someone living in the same municipality saying that one came from a family considered good in terms of moeurs, and having proclaimed its hatred for kings. This was rapidly withdrawn, as early as 1796, but the idea that a meritocracy enabled the renewal of the establishment remained.

Since 1950, and still today, the problem is different since the elites have completely adopted this system, and since the meritocracy is drawn from within the elite itself: the number of *Polytechnicians* or *Normalians* coming from the establishment is substantial, and it is not only a matter of money, it is because they have been trained for this. This issue is, therefore, connected to cultural assets and their reproduction.

Sylvestre Huet: In other words, mathematics used to be a tool for democratization and they are today a tool for reproduction, is that what you are claiming?

Jean Dhombres: It suffices to observe that the strategy of families belonging to the French establishment—which includes university professors and a certain type of engineers—and trying to maintain a certain social status, or to raise it, is to preserve a status quo regarding the essentially mathematical programs of the preparatory classes to the *Grandes Écoles*, and even more regarding its purpose of being a privileged system (by the small number of professors chosen among the most competent for purely schoolwork, by the inclusion of preparatory classes within high schools, keeping young people old enough to vote in a state of prolonged adolescence, and hence of irresponsibility). The current system enables the injection of a small dose of talents from outside this social group, and to achieve a minimum amount of competitiveness (though it is not as hard as proclaimed since most of those who prepare for the competitive examinations are admitted in some *grande école*), but it is a far cry from the revolutionary idea which was the necessity to make way for research. All textbooks of mathematical exercises for students in these classes radiate boredom to such an extent that in general they make mathematicians who believe in the freedom of the inventive mind and in creative ingenuity uncomfortable.

Cédric Villani: I would nonetheless like to point out that this shortcoming is not specific to mathematics, and that the entire French education system suffers from this

---

[11] A corps of civil engineers. Translator's note.

tendency. Our acknowledgement of reproduction, which is very accurate, is rather an admission of the failure of our entire education system.

Jean Dhombres: A political and cultural failure is all the more surprising because teachers are well trained, and examiners for these competitive tests display an acute sense of justice. They feel responsible. But there is I think an important counterpoint: the selective side of mathematics does not stem from mathematicians themselves. I am only going to recount an anecdote. For various reasons, I had accepted to give a mathematics course for pharmacology students. When examination time came, the dean of pharmacology came to see me saying: "You know that you are the one who is going to give the marks that will enable us to select the students because you are the only one who can give a 0 or full marks". This is dreadful and wholly ideological. Basically, perhaps the other disciplines have somehow given up, being unable to select their own students, they rely on an extrinsic subject—mathematics—which is not essential for the body of knowledge of these disciplines. This initial point is in my view significant.

Gerhard Heinzmann: Indeed, this anecdote is truly dreadful.

Jean Dhombres: There is another anecdote concerning the same period that I like to recount: during the first entrance examination to the *École polytechnique*, Louis Poinsot[12] was asked an algebra question and he answered: "I never learn algebra, but I promise you I will learn". He was admitted with bottom marks—having obviously shown the examiner that he knew other things. This is not selection applied to the letter, but based on ability.

Didactics of Mathematics Called into Question

Sylvestre Huet: One of the things that strikes me nowadays is the number of primary school teachers incapable of understanding the mathematics they teach, and hence of understanding why children do not master this or that concept. As lower secondary school teachers are generally former good students in mathematics, it may be speculated that they find it hard to understand mental processes that do not lead students towards comprehension. All these discussions are very important and international comparisons (PISA[13] or TIMMS[14] surveys) now reveal that the mathe-

---

[12]Louis Poinsot (1777–1859), French mathematician, member of the French Academy of Sciences, known for his contributions to rational mechanics.

[13]PISA (Programme for International Student Assessment), a survey of 15 year-olds carried out every three years in the 34 member states of the OECD (Organisation for Economic Co-operation and Development) and in partner countries; it tests the ability of students to apply the knowledge learnt at school to real life.

[14]TIMMS (Trends in International Mathematics and Science Study), an international survey published every four years, comparing mathematics and science teaching in lower secondary schools.

matical training of children is poor and getting worse. Besides, as seen above, present mathematical education contribute to social reproduction rather than to republican meritocracy. In your opinion, what should be done?

Jean Dhombres: The second point at issue in mathematics education is indeed related to the didactics of mathematics, which has diverted from its initial purpose which was trying to understand the cognitive processes at the heart of mathematics education. Since didacticians were empowered—or took over power—, to say what should be taught in mathematics, and how it should be taught, we have been heading towards an excessively normative teaching methodology ("this is how it should be taught and not in any other way"). I was on the thesis jury of many a didactician who later were in charge, and I noticed that several of them had become normative—I would not dare say "narrow-minded". Whereas every mathematician knows very well that when he teaches, reading out his notes from the first to the last line is not sufficient.

Cédric Villani: And beyond the field of mathematics, to establish a good contact with a class, a professor needs some amount of space in his teaching methodology, he needs to find his bearings and come to grips with the subject and the pedagogical method. One of the lessons from these international comparisons is the importance of the degree of freedom accorded to teachers.

Jean Dhombres: He also needs to come to grips with his class, leaving his students some leeway, though it is not easy... Nowadays secondary level mathematics syllabi are extremely prescriptive.

Cédric Villani: As far as I am concerned, of that there is no doubt: teachers that have made a lasting impression on me are precisely those who did what they were not supposed to do! I remember learning the principle of proof using barycenters at 15, and among things I learnt at school, this is one of the techniques I found most fascinating. We used to construct trees to solve geometric problems using barycentric coordinates... I had never done that before, nor have I since, but it has left a lasting impression.

Jean Dhombres: I had the same experience in my high school with barycenters in 1957 (and the same impression of seeing something tremendous with which one could solve in almost one go Euler's line or other properties of the triangle). The fact that this was still being taught long after proves that the reform of modern mathematics has not scaled down everything, as is too often claimed, and that many teachers rely on their wisdom about what furthers a taste for mathematics.

Cédric Villani: In my view, dialog between teachers should be much more encouraged, exchanges on what works and what doesn't made possible. I read the report on mathematics education for Unesco, which was coordinated by Michèle Artigue, and which analyzes this mechanism in Japan where teachers meet in some sort of seminars, act out lessons before each other, a bit like in an advanced teaching degree class; they are with colleagues, they meet each year, talk about their experience, of what worked or did not work... A horizontal mode of progression which in practice seems to give far more efficient results than a vertical mode of transmission, from inspectors to teachers or from commissions to teachers.

Sylvestre Huet: Time is another severe constraint in teaching in France, in mathematics as well as in other subjects. If teachers do no have enough time to explain the syllabus and do exercises, should the curriculum not be reduced?

Cédric Villani: To start with, lost teaching hours should be regained. The time of exposure to mathematics lost over the last fifteen years amounts to about one year.

Gerhard Heinzmann: Time is indeed a key factor, which operates at two levels. On the one hand, as class time is reduced, the weaker students fall behind all the more quickly. Exercises are given as homework, corrections are often hastily done... It all goes far too fast. On the other hand, time is used as a selection tool in examinations. However, the best are not necessarily those who have answered more questions than others within a given time. This aspect of the French education system is open to criticism.

Cédric Villani: What I am going to say may seem somewhat harsh, but mathematics is a field where it is very hard to achieve a deep understanding alone, outside the school framework. If you find economics fascinating, once the course has ended, you can digest a heap of books on economy; if you find history fascinating, you can become reasonably cultivated in this direction. By contrast, if you find mathematics fascinating, once the school course has ended, it will be very hard to go beyond something superficial, because it is impossible to do the exercises alone, which nonetheless proves to be essential. The issue of exercises as well as that of language acquisition is very specific to mathematics.

Pierre Cartier: May I contradict you, given my personal experience? I studied in a third grade high school in a city ruined by war. Nonetheless, at 15 or 16, I learnt differential and integral calculus all by myself, enough of it at least to be able to construct Bernoulli numbers (and the Euler–MacLaurin summation formula) and master the calculus of variations to recover the calculation of a ray of light in a medium of varying index, using Fermat's principle. By 13, I had understood the essential part of algorithms and trigonometric functions (Euler's book of 1748!). I also had exercise books, all of which I completed. Later, in preparatory class, I read all by myself Chevalley and Weyl in the *Sainte-Geneviève* library,[15] as none of my (excellent) teachers has heard of them. It is, therefore, possible to be self-taught in mathematics!

---

[15]The Sorbonne library. Translator's note.

Jean Dhombres: Here, it may be supposed that the free use of the internet, both by teachers and pupils, could foster the necessary training provided by exercises in mathematics. What I don't quite see, probably because of a lack of imagination and experience in secondary school teaching, is how to suggest exercises that would not just be model exercises, only aimed at answering usual examination questions.

Cédric Villani: I think the main issue in education is not so much that of curriculum, but that of organization. And regarding this, I will take a management perspective. In my view, the basic problem is that of "governance", to use a fashionable word: the system as a whole does not really have a pilot, and local aspects of interactions between teachers have not received the attention they deserve. Of course, teaching is based on curricula, but the dynamics of the school, the understanding between teachers, the right relationship with the headmaster... All these aspects play as important a role as the curricula that have been elaborated.

Pierre Cartier: This is what is called the educational community.

Cédric Villani: Exactly, and in fact the educational community does not work. There is no solidarity within the group, neither any team spirit in the good sense of the word. A secondary school is like a company that needs to be run. Who to recruit? How to integrated new arrivals? How to share tasks? Both the school system and the education system are far from being perfect and the constraints they face are immense and many, but awareness among school staff as a whole must increase that if it does not work, if it is hard, it is not the fault of flawed school curricula alone. It is also due to the way things are organized locally, to the awareness of one's responsibilities: responsibility of the headmaster towards society and his teachers whom he must defend, solidarity between teachers, coherent discourse towards students and families—in the same way that the fate of society also depends on what happens within families and the way parents educate their children based on a responsible, tolerant discourse and what not. This discussion cannot be avoided if we want an education system that works. Today all protagonists shirk their responsibilities: teachers are taken for a ride in a condescending point system, initially conceived to ensure fairness, but which today has become corrupted in practice, and the role of trade unions, also initially conceived to ensure fairness, all too often acts as a paralyzing factor of the system. As for secondary school headmasters, they are major protagonists in our society. They have a moral, educational role, which is absolutely fundamental, but this is rarely talked about even though this should be a major element in the public debate.

Gerhard Heinzmann: In the United States, headmasters are trained in a different way and school structures are built differently. The former study a special program at university and obtain a special degree. This type of appointments to head a school certainly has its inconvenience (the management aspect is likely to be very strong as well as the separation from the teaching body), but in my experience, I found that headmasters are rather well trained and one may for example consider a dual management, as is the case in American universities.

Jean Dhombres: A key point of this discussion of school management, but which is equally relevant for universities and *grandes écoles*, is to make the aims of a good management clearer. Liberal thinking which contaminates all discussions sets the debate in the sphere of competitiveness, which results in rankings, some sort of "Michelin Guide" of school learning, but we have got no clue as to what is being measured in this way. The school is not a matter of choice, but an obligation, and, to remain in the strict framework of our discussion, mathematics is not an option there.

At the end of the French Revolution, Stendhal wrote down his abhorrence of having to endure a class, promiscuity, the judgement of others.[16] Except for mathematics, as he thought that one could progress in this subject only through the common life of the class, and that an essential point was that the class should display its talents in town. But politicians have a hard time providing visibility for schools in the life of the city, and administrative authorities, educational ones at the regional level for example, are vigilant watchdogs against headmasters taking positions publicly.

Is There a Mathematical Culture?

Sylvestre Huet: Is it possible to define the basic mathematical culture a secondary school student should have? For example, we often have to deal with graphs representing the evolution of a variable over time (in other words, elementary functions). Yes, public debates or opinion polls show that these lower secondary level basic mathematical concepts are not well mastered; people do not picture what they mean simply because they have no mastery over the mathematical tool behind them. In your opinion, on what should the focus be?

Cédric Villani: The question is tricky... Let us first of all ask ourselves what we mean by "culture". Culture can cover many different things. In some cases, "culture" is the background which enables us to understand this or that concept. But what does it mean to "understand"? Is it having an idea of what it is? Is it being able to redo the calculation? In other cases, "culture" falls within the realm of wonderment, of interest, or even of fascination. And these two aspects are important and totally different. So in my view, culture is both a mind-set and a way of seeing things. For example, among all the talks I have given in very diverse social contexts, the one where I had the feeling that mathematics was perceived with greater fascination and enthusiasm was the one I gave at the festival organized by *Lutte Ouvrière*[17] in 2011 on "Mathematics and Progress". The audience displayed a fascination, an exceptional respect for the scientist I symbolized.

---

[16]He described this school experience in *The life of Henry Brulard*.

[17]An extreme left French political Party. Translator's note.

On the other hand, there is the issue of providing citizens with tools, with a foundation to enable them to understand. In this regard, probability and statistics are essential because they are used everyday in decision-making processes—when trying to come to a decision, we try to gauge the likelihood of this or that event... I recall a particular meeting with Abdoulaye Wade, the President of Senegal, who is very proud of his mathematical culture—he was initially trained as a scientist— where he explained that probability is possibly what is most useful to him on a daily basis in decision-making.

Then there is the issue of statistics, which is becoming more and more pervading because of the increasing number of opinion polls—in particular in France, they happen everywhere! What does a poll mean? What is the error of margin of a poll? How does the problem of hard mathematics, namely the understanding of statistical phenomena, interact with the human problem, namely the understanding of the fact that the answers are not always sincere, or sincerely wrong, or that samples are badly prepared so that the mathematical model no longer reflects reality... It is important to analyze all these things: polls appear everyday in newspapers, but do they have any value? Probability and statistics are perhaps the most significant elements in today's world, those that raise most questions. And yet, these are subtle concepts. It is very hard for a teacher to give a course in statistics and abide by the rules of a mathematical course based on the principle "we prove everything". I almost wonder whether it would not be better if statistics were taught as part of a different course, for example an economics course, possibly with special sessions providing a dialog between mathematics and economics teachers, because pupils must be made aware of the fact that statistics is not merely a matter of formulas or incantations, but that there is a construction behind.

Gerhard Heinzmann: Indeed, there are initial conditions that have to be satisfied, but which rarely are. Hence there is initially a sham mathematization which gives the impression that the entire thing is sound.

Cédric Villani: It is true for opinion polls, but even more true in the fields of finance, or climate.

Jean Dhombres: There are some essential things that should become very early part of the mathematical background of students, and which unfortunately are not taken account of in school curricula. For example, I am rather surprised to see that percentages are broached in the penultimate high school year in the science section in a chapter which first reviews briefly what fractions are. In other words, percentages are considered to be more difficult than fractions.

Sylvestre Huet: In general, someone who has not come to grips with fractions in lower secondary school, has problems with percentages! And experience shows that this is what happens.

Jean Dhombres: Basic mathematical culture should therefore include, I dare not say fractions, but at least percentages, specifically those constantly encountered in probability, aforementioned as an essential culture. This should be required when starting higher secondary education. But it seems to me there is something more behind this issue of percentages. For if all French people who fill in their tax forms know that a reduction of 20 %, then of 10 %, is not the same as a straightforward reduction of 30 %, few are willing to recognize that their knowledge is not merely of a practical nature, and that they are able to carry out the reasoning expressing this difference. But there is some sort of fear of voicing this.[18]

Sylvestre Huet: I would like to share with you three experiences as a journalist where I faced the challenge of writing about topics involving mathematics. The first topic: industrial risks and especially nuclear ones. When the accident of the Fukushima nuclear plant occurred, I came to write several articles conveying the seriousness of the number of Becquerels and millisieverts measured in the environment or in agriculture products from the biophysical perspective, in terms of additional risks of developing cancer during one's life, while making it clear that our knowledge on these questions is not absolute despite decades of experience of the matter. In such an article, where one is moreover not allowed to use mathematics, it is hard to explain the concept of very low doses of relative risk. For example, evacuated zones are less affected by radioactivity from the plant; the risk of death (through shock, stress, etc.) of the elderly who were evacuated in urgency on stretchers was higher than the risk of cancer from a dose of five or ten millisieverts. However, I have often checked that such an analysis is "not accepted", because risk assessment depends on qualitative rather than quantitative factors such as the origin or the nature of the risk.

Second example: climactic changes. Climatologists tell us what might happen if we release so many quantities of greenhouse gasses and if their level in the atmosphere is this or that. Given what they know about climactic machinery and about how they can be computer simulated—the only technique available to address the problem—, they obtain a bell-shaped curve about climactic reactions to a given manmade perturbation; but as they do not have sufficient knowledge about how climate mechanisms work, they cannot provide any guarantee that the midpoint of this curve indeed represents what is going to happen. Now, in the case of a very strong reaction—unlikely, in the "tail" of the curve on the right on a graph—, the risk level is so high that it justifies making far greater efforts to avoid this—because it is really worth avoiding. This argument is familiar to mathematicians, who will remain alert to the idea that "unlikely" does not mean "will not happen".

Third example: particle physics. Before the discovery of the famous Higgs boson in July 2012, when I was discussing with Cern physicists, they told me: "We are at 2, $2\,\sigma$!", which means absolutely nothing for the general public. So I asked them to tell

---

[18]In my seminars at the EHESS (*École des études supérieures en sciences sociales*) I have often addressed the general topic of the fear of mathematics, a very vague or sometime almost metaphysical fear. This is a topic in the cultural sociology of history which we do not know how to properly tackle, and which cannot be reduced to cognitivism or didactics.

me how I could explain to people who were neither mathematicians nor theoretical physicists or particle physicists, the meaning of the result in simple terms. Finally, together we found a formulation explaining this strange concept: "We will tell you the Higgs boson has been found the day when we will have a standard deviation similar to those in past experiments of the same type (where a new particle appears), for which we never had to rescind the announcement of the discovery." In their jargon, this means not 2, 2σ, but rather 4 to 6σ. When the discovery of a boson "having the characteristics of the Higgs boson" was announced in July 2012 in Cern, physicists could rely on a standard deviation of 4, 9σ, which for them was sufficiently large for any subsequent disappointment. But in regard to this as well, the number meant nothing for the vast majority of my readers. Based on these examples, what do you think should be done to increase our common mathematical awareness? You very correctly said that what are most used are probability and statistics, especially in political debates; however, it is not the easiest thing to explain.

Cédric Villani: The common point between your examples is the concept of probability and risks. The reply I would, therefore, like to give on the spot, but which will then have to be refined, is that the most important is to make pupils familiar with somewhat subtle arguments, especially involving probability. And for this, these arguments have to concern simple counting problems, or the analysis of experiments, etc.

Jean Dhombres: The teaching of modern mathematics is probably not very well adapted to the teaching of probability. There is a need to innovate, or at least to make an effort, because these are mathematics in which the subtlety of arguments is of a different level from that of logical sequences encountered by high school students. Probability is undoubtedly part of mathematics, but it is useful to show the mistakes one could be led to make. Whereas in practice, general mathematical teaching aims to avoid highlighting mistakes, lest they spread.

Cédric Villani: I have great faith in exercises. I have this idea that during a mathematics class, understanding mostly takes place when doing exercises. It is then that it all clicks into place, that one fully understands what is written in the course. Obviously, I seize this opportunity to say that—like many of my colleagues—I have signed a petition demanding more science teachers,[19] because I think that secondary level science teaching is in a very bad shape at the moment. When you look at the demands, one of the main stumbling block concerns timetables, which do not give the time to complete and give a solid foundation to the content of the course through exercises.

Returning to the previous examples, specific answers to these questions will not arise in the course. In contrast, analyzing this type of situations through exercises would not be a waste of time!

---

[19]France need more scientists: http://irem.univ-lille1.fr/PetitionLycee.

Pierre Cartier: We have discussed mathematical culture, but a parallel may be drawn with musical culture. What is "musical culture"? Everyone knows that music does not reduce to theoretical learning, but that long hours of practice on some instrument are required.

Jean Dhombres: I recall that when he was Education minister (1984–1986), Jean-Pierre Chevènement had made the surprising proposal of replacing mathematics by the history of mathematics for final year high school students in the literature section! Which obviously is meaningless. Except if a method is invented enabling us to combine mathematics with history.

Cédric Villani: In some sense, literature students are actually the key to the problem for they subsequently make up the large majority of school teachers. This fact should not be disregarded: paradoxically, our children's first contact with mathematics is mostly from teachers with a literary background. The issue of mathematical training for literature students is, therefore, fundamental for our society as a whole.

Gerhard Heinzmann: Unfortunately, the latest reform (under the Fillon government[20]) runs contrary to these aims, since mathematics has become optional for final year high school students in the literary section.

Sylvestre Huet: Same as history has become optional in scientific sections in the final year.

Pierre Cartier: Which is just as absurd!

Cédric Villani: I am sorry to go against the view of this eminent assembly: that history may be optional in scientific sections in the final year is in my view less absurd in the sense that history and geography are subjects that can only be appreciated after entering working life, the world, and after having traveled. All that one learns at high school is theoretical and disembodied. I remember my geography classes being terrible, but my way of looking at this subject has radically changed since I started traveling.

Gerhard Heinzmann: Contradicting a personal experience is impossible, but all that one learns in high school in mathematics is equally theoretical and disembodied.

Jean Dhombres: Perhaps one of the basic aspects of education in France is that pupils are no longer considered somewhat autonomous persons, who can be requested to do a certain amount of personal work. I fear that in the curriculum, he is regarded as a mere container.

---

[20]Government led by Prime Minister François Fillon from 2007 to 2012. Translator's note.

Is Applied Mathematics the Foundation of Mathematical Education and Culture?

Sylvestre Huet: To try to bring to an end our discussion on the broad topic of mathematics education, what can be taught on the nature of mathematics and its relationship with the other sciences and social activities beyond the specification of curricula or didactics? Mathematics is clearly more and more used in all social activities, whether in industry (modeling...), politics (polls...), humanities and social sciences (management tools...). What should the implications be for the teaching of this subject?

Cédric Villani: I think it is not something that can be taught in a theoretical manner; if the teachers give a class on the relationship between mathematics, the real world and society, it won't go down well. It is necessary to start from examples, and show how mathematics can be used here and there. Pierre mentioned matrix calculus earlier, and an application of matrices to a small problem in population development in final high school years is perfectly feasible. It does not appear unreasonable to me.

Pierre Cartier: In the 1960s, I gave a class on population development in terms of recruitment strategies—the technique used was of course that of matrices. I came up with ad hoc examples by more or less showing that recruitment should be according to society's age structure, otherwise we end up with "bulges" that move about, and last for a long time in case of extensive recruitment. In fact that is what happened in the 1960s with extensive academic recruitment and we are still seeing its aftereffects. And so, to answer your question, I would say that the relationship mathematics-society can be taught through its historical aspects.

Gerhard Heinzmann: In my view, one should try—I don't know whether it works—to begin from the fact that mathematics are essentially a tool—also a language, but it can be argued that language is a tool. Then the mathematical curriculum could be adapted to the physics curriculum, but also to that of economics, philosophy... Why not get different subjects teachers together so they define a corpus of knowledge adapted to what each can use as mathematics in his class?

Jean Dhombres: In this respect, I would remind you of the huge failure of the TICE program (Information and Communication technologies for teaching) launched in the 1990s in high schools, and which amounted to making pupils work on subjects in which mathematics is applied in various forms. And at the request of mathematics teachers themselves this program was abolished—not to mention trade unions' objections about means, etc., which are understandable. Their main argument was: "We, mathematics teachers, do not have the skills to help other subjects". I recall a very competent mathematics teacher explaining that he was for example incapable of explaining what the lift action on an airplane was using mathematical elements.

Cédric Villani: Indeed, because we need to find the degree of simplification...

Gerhard Heinzmann: But this is what academic institutes for the training of school-masters were supposed to be for.

Jean Dhombres: We all agree. And their second argument was: "Mathematics is a training in intellectual rigour, and all we would do in applications would lack rigour". Of course, this is all ideological, but the teachers' side also needs to be considered.

Cédric Villani: It is mainly a question of teacher training. I think this issue of collaboration is important and in this regard much effort is required. At the same time, the reasons given are not trifling. they cannot be brushed aside. It is true that mathematics teachers are not trained to intervene in other subjects, and their honesty in saying so ought to be recognized. In consequence, new generations have to be trained. In other words, this is going to play out over the next forty years—that is the time needed to replace a generation of teachers. Which does not rule out the urgency of implement the training.

Pierre Cartier: Is the single-subject teacher not the original sin in the French system?

Gerhard Heinzmann: I didn't dare say so.

Cédric Villani: It is a delicate issue, because it was seen as a progress with respect to the status of lower secondary teachers, who were trained to teach two subjects.

Pierre Cartier: In practice—and I am guided here by my mother's experience as a high school headmistress—it was a way of bringing primary school teachers into lower secondary teaching on a massive scale, at a time when the latter was rapidly developing, and the section "superior primary education" was being abolished—a first step towards a common lower secondary school for all. These multi-disciplinary teachers all too often did not master any of their two specialized subjects. And in terms of staff management, this created extensive problems. Before 1950, single-discipline teachers were the rule, more pronounced in the sciences than in literature (where the same teacher used to teach French, Latin and Greek, or history and geography). Let me add that, in the current situation, arguments should be qualified and general high schools should be distinguished from technical ones.

Jean Dhombres: I think the training and recruitment of multi-disciplinary teachers should be allowed—but not made compulsory. But we are Jacobites in French education: we do not want to see anyone standing out, or rather we want everyone to stand out.

Cédric Villani: It is again a matter of freedom. The elements that one usually bears in mind and that in any case follow from the PISA ranking is that education systems that work best are those where teachers have room to manoeuvre, where local decisions can be taken account of—this needing to be qualified.

Sylvestre Huet: It is indeed a complicated proposition, because the quality of teaching on an entire territory is a key factor of equality between citizens, and that any local variation will be necessarily interpreted in social terms, namely "the minimum will be done for high schools in popular suburbs and the maximum for those in the wealthiest neighbourhoods". Besides, there are numerous arguments to support this interpretation about the possibly hidden purpose behind...

Teaching the History of Mathematics

Sylvestre Huet: Would teaching the history of mathematics, and more generally, the history of science not be something powerfully efficient in cultural terms? In old conceptions of literature and social history teaching, the approach was deeply chronological. Science was profoundly grounded in this mode of perpetuation, and mathematics even more so than the other sciences. There is a cumulative aspect— which in the natural sciences was also done by subtracting, since what once understood as a truth is now understood as a mistake, an insufficiency... History of science is in my opinion a powerful tool to understand science itself. Not just this or that particular result, but the way they work, their relationship with reality and their rationality. Do you share this idea?

Jean Dhombres: My reply will be very precise. It is the one I gave Jean-Pierre Chevènement when he was Research and Technology Minister: when you teach the history of economics or the history of literature, you are doing economics, you are doing literature. Is learning the history of mathematics, doing mathematics? You cannot just say when Newton is born, or merely recount what happened, without bothering about was happened within. A historical way of doing mathematics is perhaps a solution, but nevertheless you are doing mathematics.

Pierre Cartier: Various different viewpoints have been given on the history of mathematics, and you have there an indication of the controversies that have swept over the entire field of history. *Event-driven history* is no longer well rated. The depraved behavior of the Duchess of Courlande, niece of Talleyrand, whom she followed in his exile in London during the *Restauration*,[21] perhaps influenced the political game of European powers. That the niece of Newton, a great seductress, was that mistress of the prime minister certainly played a role in his public career. But Newton, become one of the most important figures of the kingdom, had finished his scientific work, and that is *what* matters to us.

---

[21] Period following the fall of Napoleon's empire in 1815, which ended with the revolution of 1830. Translator's note.

Dieudonné[22] and Weil,[23] either under their own name or under the pen of Bourbaki, promoted a *conceptual history* of mathematics. It consists in following the long-term evolution of concepts, changes in the meaning of a word (point, line, number...) that has remained the same. It is tempting to rewrite history in the light of present certitudes, forgetting that they are not eternal.

André Weil also advocated the idea that Gauss or Euler should be considered our contemporaries, as our colleagues from across the corridor. In other words, you have to associate with classical figures. This method is well suited to European mathematics since the end of the 17th century. I followed the advice of my master Weil, and benefited much from reading Descartes, Pascal, Fermat, Newton, Euler, Lagrange, Jacobi, Riemann, Dedekind.... I found there the inspiration for my own present work. It is in this sense that learning the history of mathematics is doing mathematics!

There are many other way of learning the history of mathematics: ethnomathematics, the history of techniques and instruments, of school textbooks, of scientific institutions (such as academies...). There is also an extensive repository of Indian, Chinese, Arab mathematics, for the moment reserved for a few specialists.

As for the application to teaching, each of these directions can be useful, but I don't think there has been any thinking of a pedagogical nature on this issue.

Cédric Villani: I would like to make a comment on the cumulative issue: it is true that mathematical knowledge accumulates more, in the sense that when something is true, roughly speaking, it remains true. Of course, some mathematical concepts become useless, but in other disciplines, entire disciplines undergo complete change. In the life sciences for instance, at the advent of genetics, the entire issue of transmission became totally changed.

And on the issue of history, I quite agree with Jean. It is an extremely powerful pedagogical tool that we have all used, including today around this table, to get concepts across. Likewise, in public talks, a concept is always put in its historical place, time, and culture. For us too, in our research, history is obviously important. I am not at all a historian of science, but it is important for me to be sufficiently well versed in culture to be able to put my talks in their historical context and to also get a message across. The historical approach is a catalyst to get people to discover a field, to become interested in, to also understand the motivations of those who developed this field. In this way, integrating the history of mathematics to didactics, pedagogy, is a very efficient tool.

But it is mainly political history that is covered in history, some economic history, and hardly any social history. One should also talk about technological history and scientific history, which are the two most visible facets of our society. When a kid looks at a history book, what mainly fills him with wonder is not regime changes, but how clothes, habits, technological tools change. The main evolution in the history of

---

[22]Jean Dieudonné (1906–1992), French mathematician, founding member of the Bourbaki group.

[23]André Weil (1906–1998), French mathematician, founding member of the Bourbaki group.

humanity is the advent of electricity and cars. This is what changed our environment, our face, our way of seeing things.

Pierre Cartier: And we should not forget the discovery of longitudes in the 18th century for navigation. The entire colonial expansion was based on this.

Cédric Villani: You are right, and all these things have as much impact as regime and leadership changes. A history class with a modern perspective should somewhat include these aspects. Thus, the French Revolution is perhaps the period in the history of France when convergence of view between politicians and scientists was at its highest.

Gerhard Heinzmann: Personally, I am committed to the introduction of the history of mathematics, not at school level, but at least at university level. I consider it irresponsible to make mathematicians study without giving a course on the history of mathematics, at least during one semester, so that students learn there is a methodology and special tools to do history, that history is neither a string of anecdotes nor of knowledge, but an epistemological activity (in the sense of *"erkenntnisbringend"*) which, like mathematical activity, requires setting perspectives.

Jean Dhombres: I agree with Gerhard. History of science, and in particular history of mathematics, should be compulsory in the training of university professors. And I think that in France there is a network of sufficiently many competent people in the history of mathematics so there is no fear that making this teaching compulsory ends up infringing the commitment to equal distribution throughout the territory.

Cédric Villani: It also depends on individual cultures. I remember the radical change in my perception of history of science when I arrived at the *École normale supérieure* in Lyon, where my neighbor was Étienne Ghys,[24] who placed great emphasis on historical aspects. For me, this was somewhat revolutionary, and I think it is truly fundamental.

Gerhard Heinzmann: I am not a professional mathematician, but I can well imagine that for a researcher it is important to see how a problem was developed, the mistakes, the wanderings, the difficulties in developing a concept...

Cédric Villani: And then also why a certain definition and not another was taken, the motivations that led to this choice, etc.

Pierre Cartier: It should be mentioned that this lack of interest for the history of science is far from being a specifically French failing...

---

[24]Étienne Ghys, French mathematician born in 1954, CNRS research director in the pure and applied mathematics department of the *École normale supérieure* in Lyon, member of the French Academy of sciences since 2005.

Gerhard Heinzmann: On the contrary, I find that history of science is quite developed in France, far more than in other countries.

Jean Dhombres: There is another phenomenon that seems interesting to me: history of mathematics is a field where the person who is learning mathematics can take a critical approach. For a high school pupil it is not easy to take a critical approach. But throughout the history of mathematics, one sees mistakes, for history is also made up of mistakes, and it is extremely instructive. The fact that great mathematicians made mistakes takes away the sacred aspect, and gives the learner the possibility to talk about it in the framework of a critical exchange. While it is difficult for a pupil to talk about mathematics—except to say "I have understood" or "I have not understood"—, a critical exchange enables him to initiate a dialog and so to appropriate the language and the underlying concepts. In this sense, history can help, provided it has at least some content.

Cédric Villani: We could broaden the issue of the role of history of mathematics to the role of scientific culture. Should scientific culture be considered something as somewhat isolated, or as an integral part of culture? It seems obvious to me that science is part of our lives, and hence also a part of our culture.

Jean Dhombres: As Cédric just said, history of science has the merit of presenting the motivations of those who made science, as well as highlighting the fact that science does not happen automatically, that there are struggles, even enemies...

The Nature and Challenges of Mathematics Research

The Mixed Heritage of the Bourbaki Paradigm

Sylvestre Huet: Now I would like us to address aspects connected with the nature and the challenges of mathematics research. Today, what are its frontlines, its relationship with scientific research in other fields, natural sciences as well as social sciences, including—we will probably talk about this—economy and finance with its all too well-known consequences? We could perhaps begin by defining the contents of mathematics research, topics on which mathematicians work, how this research is organized, within what structures…?

Pierre Cartier: I would say that we are in a period where we are reaping the benefits of the years 1950–1970, which were some of the most exciting ones—at least for French mathematics—since this was the golden era of Bourbaki. Personally, I never sought to hide the fact that I was part of this adventure, contrary to those who think these intimate secrets should not be revealed. We talked earlier of the changes in mathematics education during this period, and indeed, I never twice gave the same class when I used to lecture in Strasbourg from 1960 to 1970. All had to be invented, it was really exciting. At the same time, it was Alexander Grothendiek's[1] peak period, who undeniably ruled over the *Institut des hautes études scientifiques* during these same years (1960–1969). It is a rather incredible period when it is fair to say that each day was accompanied by something new: Jean-Pierre Serre's[2] thesis in 1951, Claude Chevalley's[3] great symposium series from 1956 to 1959, which I collaborated

---

[1] Alexander Grothendiek, stateless mathematician born in 1928 living in France who received the Fields Medal in 1966, and is considered to be one the greatest 20th century mathematicians.

[2] Jean-Pierre Serre, French mathematician born in 1926 who received the Fields Medal in 1954 and the first Abel prize in 2003, and is considered to be one of the most important 20th century mathematicians.

[3] Claude Chevalley (1909–1984), French mathematician, founding member of the Bourbaki group.

© Springer India 2016
P. Cartier et al., *Freedom in Mathematics*,
DOI 10.1007/978-81-322-2788-5_4

on and reedited later on, Grothendieck's early works... To attest to the quick pace of progress: in 1956–1957 it was still somewhat hesitantly that algebraic groups were mentioned; in his 1958 talk at the international Congress of Mathematicians in Edinburgh (ICM), Serre said: "I am going to talk about algebraic geometry, that is to say about schemes", whereas two year earlier "schemes" had never yet been mentioned. It is from a fairly French standpoint that I want to recall this dynamic period, with the advent of topology, of complex functions, of algebraic geometry... When I reread the seminars of the time, I am stunned to understand them, when I used to find them so hard in those days! The twenty years from 1950 to 1970 completely changed the mathematical landscape. And this change has a name: "Bourbaki". This is too general a statement, but let us say that most of the protagonists of this change were actually members of Bourbaki or very close to members of the group. There was continual interaction: internal meetings within Bourbaki prolonged discussions held in public, and public discussions prolonged discussions held internally. There was indeed a paradigm shift—if the expression "paradigm shift" has a meaning.

These years also coincided with the Algerian war (1954–1962), and those who used to meet on some evenings at Laurent Schwartz's[4] house were the exact same who attended Bourbaki meetings. We all had a foot in politics and a foot in mathematics—with differences of opinion obviously.

During these years, mathematics completely changed, and not only in the field of algebraic geometry. Probability theory, to which I then contributed, also underwent radical change, in particular thanks to the work of Jacques Neveu[5] and Paul-André Meyer[6] in France, of Joseph Leo Doob[7] in the United States...

Cédric Villani: Yet, neither Jacques Neveu nor Paul-André Meyer belonged to Bourbaki.

Pierre Cartier: Meyer, partly. He attended some meetings and helped write the only Bourbaki book where Brownian motions are mentioned.

But this paradigm shift did not merely take place in France. As David Mumford[8] told me one day: "On the one side, we [The Zarisky[9] school in the United States] had the problems and had somehow reached the end of our methods, and on the other side there was Grothendieck who had methods but no problems!" All that

---

[4] Laurent Schwartz (1915–2002), French mathematician, first Frenchman to receive the Fields Medal in 1950, and member of Bourbaki. As an engaged intellectual, he distinguished himself by his numerous political battles, especially against the Algerian war.

[5] Jacques Neveu, French mathematician, one of the founders of the French school of probability.

[6] Paul-André Meyer (1934–2003), French mathematician, one of the founders of the French school of probability.

[7] Joseph Leo Doob (1910–2004), American mathematician, known as one of the founders of the theory of martingales (probability theory).

[8] David Mumford, American mathematician born in 1937, student of Oscar Zariski, known for his work in algebraic geometry. He is now interested in vision and image recognition problems.

[9] Oscar Zariski (1899–1986), American mathematician of Russian origin, whose contribution to the development of modern algebraic geometry is major.

was needed was shaking hands… With great generosity, Zariski sent Grothendieck his best students and he himself gave a talk on Serre and Grothendieck's work on coherent sheaves, etc. This small Franco–American "miracle" took place, but it is only later that we realized that Grothendieck's best disciples are neither American nor French, but Russian. As can be seen today, the most faithful to his adventurous spirit include Maxim Kontsevich,[10] Yuri Manin,[11] and others. And the other "miracle" is the communication that occurred with the Soviet school, which could not be taken for granted by any means. Each time Bourbaki published a volume, Manin or someone else used to ask us if they could translate it into Russian.

Cédric Villani: The history of probability in the 20th century is from the very beginning primarily Franco–Russian: Émile Borel,[12] Andreï Kolmogorov,[13] Paul Lévy[14]…

Pierre Cartier: Americans should also be added.

Cédric Villani: Its true there was Doob in the United States and Itô[15] in Japan, so let us say that it is Franco–Americano–Russo–Japanese!

Sylvestre Huet: You therefore claim that mathematics research underwent a paradigm shift in the years 1950–1970 and that since then it has only consisted in work based on the ideas of this new paradigm.

Pierre Cartier: Yes, and this is, moreover, what transpires from the history of science: major paradigm shifts are followed by stable scientific periods, when the new ideas are exploited to the full. It is a rough outline, but which explains a good many things.

Cédric Villani: Well I am going take the opposing view, merely because I belong to a different generation—and even if the above point of view and mine can probably be reconciled. I belong to a generation for whom Bourbaki is a pejorative word. When a piece of work is labeled "Bourbakist", in the sense the word is given today, it is

---

[10]Maxim Kontsevich, Russian mathematician born in 1964 who received the Fields Medal in 1998, tenured professor at the *Institut des hautes études scientifiques* (France).

[11]Yuri Manin, Russian–German mathematician born in 1937, known for his work in algebraic geometry and in diophantine geometry.

[12]Émile Borel (1871–1956), French mathematician, politician and philosopher, specialist of the theory of functions and of probability, member of the French *Académie des sciences*.

[13]Andreï Kolmogorov (1903–1987), Russian mathematician, whose name is especially associated to probability theory, dynamical systems, information theory, and topology. He is considered to be one of the most important mathematicians of the 20th century.

[14]Paul Lévy (1886–1971), French mathematician, one of the founders of modern probability theory.

[15]Kiyoshi Itô (1915–2008), Japanese mathematician, whose name is associated to probability theory and to a fundamental lemma in stochastic calculus (Itô's Lemma).

never a compliment! I nonetheless am one of those who consider that the Bourbaki heritage is fundamental.

Indeed Bourbaki led a revolution, that I would compare to the excitement that affected physics in the early 20th century (1900–1930), with the emergence of quantum mechanics—"the thirty years that shook physics", as George Gamow[16] used to say. The years 1950–1970, Bourbaki's golden age, with its redevelopment, its feeling of rediscovering everything, were followed by twenty years when mathematics underwent a complete change, but with a radicalization of the mathematical discourse with respect to other scientific disciplines. But, in my opinion, there has since then been another revolution, like that led by Bourbaki, but concerning how it is integrated to the rest, and which corresponds to the fall of the wall separating "pure" mathematics from "applied" mathematics.

Pierre Cartier: It is true that since about 1990, we have entered a new phase of turbulence, of developments in all directions. Regarding the relationship between "pure" and "applied" mathematics, this is one of the points where I differ from those who accompanied me in those years.

Cédric Villani: Indeed, on this point, you are not at all typical. But it is during that period (roughly speaking in the years 1940–1960) that a real antagonism arose—which did not exist earlier and which does not exist now—concerning the relationship between mathematics and physics and especially between pure mathematics and applied mathematics. People used to refer to themselves as pure mathematicians or applied mathematicians, and this influenced their opinion. When my generation hears these people give a talk presenting themselves as a pure or applied mathematician, it gets the impression of hearing an outdated debate. For example, when Pierre-Louis Lions,[17] who was my thesis supervisor, received the Fields Medal in 1994, he said—deliberately or not: "I am the first applied mathematician to received the Fields Medal." Outside France, Lions' work would not be regarded as being applied mathematics, but pure mathematics—and an Anglo-Saxon would laugh.

In contrast, at the last International Congress of Mathematics in 2010, we were two from France to receive the Fields Medal (Ngô Bao Chau and myself), one Russian (Stanislav Smirnov) and an Israeli (Elon Lindenstrauss), and during the first press conference the unavoidable question was asked: "Pure/applied mathematics, what do you think?" The joint answer of all the medalists: "It is a distinction that used to be, that no longer has any meaning, mathematics is one whole, pure/applied the same battle," The tune is radically different from the dominant one thirty years ago. There is a genuine evolution which amounts to an assimilation, and which also

---

[16]George Gamow (1904–1968), American theoretical physicist and cosmologist of Russian origin, whose name is associated to the Big Bang theory. Apart from scientific works, Gamow wrote several popular science books, in particular the Mr. Tompkins series.

[17]Pierre-Louis Lions, French mathematician born in 1956 who received the Fields Medal in 1994 for his work in the field of partial differential equations. Member of the French Academy of sciences (1994), he is a professor at the *Collège de France* since 2002, where he holds the "Partial differential equations and applications" chair.

corresponds to the constantly increasing importance of probability since the start of the 20th century.

Pierre Cartier: But remember that you and Werner[18] are in some sense, the first probabilists to received the Fields Medal.

Cédric Villani: In some sense, yes. And Smirnov as well, since we are both probabilists-analysts. But it is in fact Werner who was the first probabilist to received the Fields Medal. It was a long awaited sign, a big change, a recognition and a mind-set as well which sees mathematics as being more central to society. Probability theory is indeed caught up with the real world, network questions, the swift development of computer science—theoretical as well as applied—the development of the internet... All this is part of a whole where applied aspects of mathematics and its interactions take a much more concrete form and are part of a change of mind-set, even a change in the aesthetics related to computer science. We are more constructive today than twenty years ago. The axiom of choice enables us to choose transcendentally, as if by magic, an element from each set of a collection, this is absolutely not constructive; youngsters like me want to get rid of it, whereas Bourbaki has no issue with it.

Pierre Cartier: I think there is some optical illusion here. In my generation too, the axiom of choice was far from being uncontroversial, van der Waerden[19] was reluctant to use it is his book *Modern Algebra* and Henri Cartan was proud that his proof of the existence of Haar measure did not use it. I would rather say that we were resigned to admitting it. As for Deligne,[20] he clearly admitted that he uses the axiom of choice out of convenience, but that he does not believe in it. I don't know whether he wrote this down anywhere, but he often said it.

Cédric Villani: I didn't know, that's interesting... In my view, there is a more recent revolution, not so much in the corpus nor in the foundations, but in the integration of mathematics with computer science, biology, physics, and through joining pure and applied mathematics that once again are going hand-in-hand.

Jean Dhombres: This fascinating dialogue shows the difficulty faced by a historian when trying, not to juxtapose, but to associate the protagonists' feelings to what is provided by the assessment of mathematics in various countries since 1950 for example, and by statistics based on *Mathematical Reviews*, this monthly journal where all mathematical articles published throughout the world are criticized, and recorded according to a rather sophisticated subject classification. Functional analysis

---

[18]Wendelin Werner, French mathematician born in 1968 who received the Fields Medal in 2006 and whose work focuses in particular on probabilistic phenomena (Brownian motion, etc.).

[19]Bartel Leendert van der Waerden (1903–1996), Dutch mathematician who wrote a fundamental book, *Modern Algebra* (1930), and revolutionized this field of mathematics.

[20]Pierre Deligne, Belgian mathematician born in 1944 who received the Fields Medal in 1978, Professor at the Institute for Advanced Study in Princeton.

and the halting of the Banach program due to Enflo's proof in 1972, Schwartz and Sobolev distribution theorems since 1944, or the convergence almost everywhere of the Fourier series of a square integrable function by Carleson, programming languages with Knut, or developments in combinatorics with Tutte, amongst others, would reveal a different picture. Not to mention the "general systemology" due to Ludwig bon Bertalanffy. The dialectics remains on-going, no longer between pure and applied mathematics where Cédric correctly explained it is now old-fashioned.

The Revolution Underway in Mathematics

Sylvestre Huet: I suggest we continue this discussion in two additional directions. First, how is this integration of mathematics reflected in terms of mathematical activity, results or still unanswered questions; for example, I guess that the problem of network stability, or security, for instance in case of the internet is a not so trivial mathematical problem. Conversely, what are the basic "pure" mathematics problems, to use an old-fashioned term, that are revived by these interactions with other sciences or arise from them?

Cédric Villani: Both questions are related.

Pierre Cartier: I shall answer by relating a revealing story. In the 1960s, for petty financial reasons, I worked as an adviser to industry at the *Compagnie générale de télégraphie sans fil* (CSF). In fact, I had to give classes to engineers. In this framework, using the field of four elements, I solved a small problem of coding/transcoding— which today has become depressingly trivial, but at the time engineers were completely bowled over. My force amounted to a conviction that modern mathematics, in particular algebra, provided new tools. What I want to express through this example is that the two revolutions alluded to previously are not independent, nor the two question you have asked, the scope of mathematics had to become wider and amplified before the scope of mathematical applications could diversify.

Cédric Villani: This is quite possible. Besides the movement you were describing, with the Bourbakist expansion, where France was at the cutting edge, does not prevent France to be at the cutting edge now with this other movement I am talking about. Experience acquired on one side is then used on the other, even if people are not necessarily aware of this.

I will answer your question by making the following comment: if you take the batch of 2010 Fields Medals, three out of the four recipients directly or indirectly work on problems coming from physics. Personally, three-fourth of the problems on which I have worked are mathematical physics problems. Smirnov too. Lindenstrauss has worked on dynamical systems which also include a physical base. This physical inspiration is omnipresent in numerous recent research works.

Pierre Cartier: One of the origins of this second revolution you are talking about is historical, and directly linked to the fall of the Soviet system, which led to the dissemination of Russian scientists throughout the world on a massive scale, in particular in mathematics: the Moscow school lost its best people. Now, Russians have never separated pure and applied mathematics.

Cédric Villani: Indeed, Russians have kept this very original feature which gave them a vision of the problems that nobody else had, and this desire not to separate theory from the operational part.

Sylvestre Huet: I understand this geopolitical explanation, but this increasing number of connections between mathematical and physical activities, not only within academia, but also within industry—as we saw with Pierre Cartier's example—, is it not simply connected to the extraordinary diversification of industry and its productions? An industry that, since the end of the 1960s, uses to the full fundamental scientific knowledge, about quantum mechanics (lasers, medical imaging, etc.) as well as about atomic nuclei (nuclear industry), etc. And this industrial movement is still growing today by assimilating in its production extremely cutting-edge knowledge coming from physics, as can be very clearly seen in computer science, chemistry, nanotechnology… Finally, is this tremendous suction process in the productive activities not the basic reason for the diversification of mathematics?

Cédric Villani: It is indeed one of the basic reasons, obviously together with the major scientific even of the second half of the 20th century: the development of computer science. Computer science has had an amplifying effect on mathematics which has permeated every field. In this regard, I strongly recommend reading "European Success Stories in Industrial Mathematics", published in 2011 by European mathematical societies, and which gives an inventory of current industrial projects with important contributions from mathematics. Behind all the examples dealt with in this book, there is clear evidence for the amplification due to computer science which enables us to change the resolution, to go from a mathematical concept to an operational level.

Pierre Cartier: Is this not merely a continuation of the American program initiated during the Second World War? The continuation of the Manhattan project? Microcomputing developed in the 1960s–1970s in a country which is allegedly the champion of free enterprise, but what did this cost American federal funds? 70 billion dollars of investment in microcomputing for the space program. In some ways, everything around us is the result of this investment.

Cédric Villani: The world is entirely mathematical, a mathematics amplified by computer science. Examples I quite like to develop in my talks naturally include optimal transport[21] on which I have worked, and I then talk of Kantorovich,[22] the pioneer of modern mathematical economy, but also of the theory of price, of optimum allocation, of planning, and then of linear programming which nowadays is one of the most widespread techniques in the world. It is used each time the need to distribute limited resources arises and that costs are proportional to the quantities involved. It is thus used nowadays by companies to plan transport charges, with thousands or millions of variables. Pierre talked earlier of the efforts needed to invert a 4 by 4 matrix in the 19th century, but today in linear programming problems, monstrous systems are inverted daily.

Gerhard Heinzmann: Just one small comment. You are witnesses of this period, but I as an observer, I see that Bourbakists went to study in Germany in the 1930s and that in fact they are the inheritors of Hilbert.[23]

Pierre Cartier: Of course. Before Bourbaki, there is Hilbert.

Gerhard Heinzmann: For me, a certain mathematical development reached its peak and ended with Bourbaki's work.

Jean Dhombres: This remark seems to me very important, and in my view, its meaning should be taken to be the same as saying that Euclid's *Elements* mark the end of the development of mathematics in the 5th and 4th Greek centuries. And just like Euclidean vocabulary has been accepted, Bourbakist vocabulary has also become commonplace globally, in particular in topology.

Gerhard Heinzmann: During my studies in Heidelberg (1971–1977), there were both an applied mathematics institute and a pure mathematics institute, and no one really knew how to deal with computer science. Applied mathematics had exhausted their methods and new ones needed to be developed. And this development of applied mathematics happened through computer science. All the logicians who were my contemporaries have become computer scientists.

Cédric Villani: Yes, logic has moved towards computer science. This is now very clear with programs being developed around computer-checked proofs, namely auto-

---

[21] Optimal transport: a mathematical field at the crossroads of analysis and probability, which can be described as the search for the best possible match between production centers and consumption centers, so as to minimize the total cost of transport of a given commodity... This operational research problem, reformulated in an abstract manner, reaches into several different branches of mathematics, like non-Euclidean geometry.

[22] Leonid Kantorovich, Russian mathematician and economist born in 1912 who received the Nobel Prize in economics in 1975. He symbolizes the duality of mathematics: both a pure art and a formidable tool for acting on reality.

[23] David Hilbert (1862–1943), German mathematician whose influence on mathematics from 1900 to 1950 was considerable. His name is associated to a number of mathematical terms and theorems (Hilbert space, Hilbert's fundamental theorem, etc.).

mated checking. This is a significant trend with major challenges, though initially it was a matter of logic, of the truth of proofs, and here too we go back to Hilbert's problems. It could then be argued that if we consider the two major figures Hilbert and Poincaré, then Bourbaki carried on Hilbert's way of thinking and the Russians recuperated Poincaré!

Gerhard Heinzmann: I don't think this is quite correct historically.

Pierre Cartier: On the whole I agree. These two giants are known to have been in competition: Hilbert's talk in Paris in 1900 on his 23 problems in mathematics was followed some years later by a talk by Poincaré on future prospects. But in France, Poincaré became wrongly associated with a certain degeneration of French mathematics, and it became necessary to react. Poincaré did not have a good reputation in Bourbaki circles of my youth, until he was rehabilitated by Dieudonné.

Jean Dhombres: Where would you place someone like John von Neumann[24] in all this; he was obviously deeply influence by Hilbert, and emigrated to the United States in 1930 and in particular developed quantum mechanics and the abstract theory of algebraic systems.

Cédric Villani: He was mostly fascinated by turbulence. I think he is one of the visionaries of computation tools who realized that to go beyond and to develop each other's work, sophisticated, automated tools would be needed.

Jean Dhombres: Can we say that the paradigm shift concerning what mathematics can do began with von Neumann in 1930? In fact, Von Neumann was invited by Weil in the 1930s to the first Bourbaki seminar on the foundations of quantum mechanics…

Pierre Cartier: To oversimplify, I would say that 20th century mathematical legacy comes from two sources: from 1900 to 1950, Hilbert and his legacy; from 1950 to 2000, computer science; and at the threshold of both periods there are people like von Neumann, Turing[25]…

Jean Dhombres: And Poincaré?

Cédric Villani: Poincaré was overshadowed during the Bourbaki period. Today he is once again in broad daylight; he is seen as a precursor with a determination to describe the world qualitatively. And moreover, now computers substantiate Poincaré's theories.

---

[24] John von Neumann (1903–1957), American mathematician of Hungarian origin, considered to be one of the most influential figures of the 20th century. Areas he worked in include game theory, set theory, artificial intelligence and computer science.

[25] Alan Turing (1912–1954), British mathematician and logician, considered to be one of the first pioneers in computer science and artificial intelligence.

The Stakes Involved in the Mathematization of Industrial Activity

Sylvestre Huet: Can you give some significant examples of questions mathematicians are asked—by large groups in the scientific and technological sector who use mathematics—and which today require progress?

Cédric Villani: First and foremost, network questions. To understand how a network works. Then problems of data transmission across a network: at what speed does data propagate on the internet, in terms of what parameters, what is it blocked by... The same problem occurs in epidemiology: disease transmission, propagation speed of a virus, in terms of what parameters, etc. Since we are discussing networks, let us talk about finance: what are the relevant physical parameters to analyze financial exchanges? What are the indicators to detect that a crisis is going to happen? Where? What are the weak links? Etc.

Gerhard Heinzmann: But mathematicians don't do this type of work.

Cédric Villani: These are, however, mathematical problems and problems lacking mathematical tools.

Gerhard Heinzmann: But the preliminary data or the choice of parameters is in no way mathematical.

Cédric Villani: Thats true, they are fixed by agreements, conventions... but it would be unreasonable to think that these will lead to our salvation!

Sylvestre Huet: Concerning finance and mathematics, I have noticed a migration of theoretical physicists who have gone from theoretical physics to applied mathematics and from there to finance, especially when Americans decided in 1993 to discontinue the Superconducting Super-Collider (SSC) project, i.e. the particle accelerator which was supposed to compete with Cern's Large Hadron Collider (LHC). These theoretical physicists went to work for Wall Street, where they developed derivative products, based on very sophisticated mathematical systems, probably incomprehensible to managing directors of banks or Finance ministers. When the crisis came, Jean-Pierre Bourguignon,[26] a mathematician that you all know, made the following comment:

> If as mathematicians, we had needed to ask questions on the reliability of these financial constructions – and in particular derivative products, etc. – to study their potential for crisis, it would have been very bothersome because basic data are kept secret and are only known by financiers.

---

[26] Jean-Pierre Bourguignon, French mathematician born in 1947, director of the *Institut des hautes études scientifiques* (Bures-sur-Yvette, France) since 1994.

The failure to share mathematical knowledge about these systems prevented the entire mathematical community from taking an interest in the problem and seeing whether this financial system could be socially problematic.

Likewise, studies in probability theory regarding industrial security, whether in the nuclear or the space sector, provided political leaders and citizens with arguments for the construction of nuclear plants or space shuttles. And these methods and their results have obviously not be correctly reviewed, at least not fully, including by the mathematical community who would perhaps have come up with computer scientists' and statisticians' favorite criticism—*garbage in, garbage out* (the results' quality depends on entry data quality): you carry out probabilistic analysis on the reliability of all the components of a nuclear plant, but how do you determine the likelihood of manmade errors or of events as improbable as those that happened in Chernobyl or Fukushima? What is the probability that a corrupt safety authority connected to industrial interests will be incapable of enforcing a genuine safety review in a place where a tsunami already occurred 100 years ago?

The stakes involved in the mathematization of industrial activities are tremendous because mathematics has been used as arguments by policy-makers, governments, business leaders and citizens. But what guarantees do citizens—or even governments—have that the mathematical community is in a position to critically analyze risk management?

To broaden the discussion a bit more, it is clear that in our societies, where electricity has become, literally speaking, a vital fluid, and where it is hard to imagine even twenty-four hours without electricity, we are facing a reliability problem with the system, which surely can be analyzed using a series of mathematical methods. How should this problem be tackled? What questions does this problem give rise to? How can citizens be given, not the guarantee, but a certain trust in the fact that the scientific community, including mathematicians, are going to tackle this kind of issues motivated solely by public interest—which is obviously not the case when an industrialist asks a mathematician to provide him with a solution to produce what he has already decided to produce?

Pierre Cartier: Generally speaking, it is very hard to obtain relevant data about problems in which mathematical analysis has a political or social impact.

Jean Dhombres: The current situation is perhaps more complicated, but it is not new. This issue of "secret" data which the public is interested in already arose in the years 1700 when the major powers of the time decided to develop an important fleet. About 10 % of warship launches failed due to metacentric causes.[27] So in France, in England, in Prussia, various scientists were asked to rectify this expensive problem—Euler was commissioned in St Petersburg, and Pierre Bouguer in France.

---

[27] On the vertical plane through the center of gravity of a ship and perpendicular to its axis, and for a given inclination, the metacentre is a point on the corresponding vertical to this inclination, passing through the center of gravity of the submerged hull. It is the characteristic point, in other words the one corresponding to the envelope of all these lines. The stability condition of the ship for a given inclination is that the corresponding metacentre be above the center of gravity of the ship.

They invented the metacentre before 1750, in other words a theory giving the static stability of a ship, using only differential geometry applied to the buoyancy curve on a ship's hull. And they faced the problem of whether or not they could disseminate their results. Despite the fact that the State, who was their sponsor, had obviously asked them to remain quiet, the scientists in question published all their results, in the name of the academic world, namely their world. In those days, a country like France or Great Britain spent about 3 % of its gross national product (GNP) on shipbuilding, which is considerable.

Cédric Villani: I very much appreciate this historical approach as too often we keep our noses to the ground in relation to problems one encounters, when in fact they recur time and time again in different corners of the world. It is, therefore, extremely interesting to see that this issue of the scientist with respect to the politician and secretiveness is nothing new. In my view, scientific aspects cannot be separated from political ones. Scientific analysis acquires meaning only if it is based on some rules, some political choices. It is meaningless to say that an economic model is validated or that there is x per cent risk if it is not accompanied by a coherent political arsenal. It is also meaningless to apply a mathematical theory outside the framework it was developed for, short of remaining aware of the risk taken by doing so. It is typically the case in finance, but also in biology (population biology, epidemiological studies…), where the mathematical assumptions within the theory are systematically overlooked. It makes sense if one is aware that the model is being applied outside its area of validity, and that this risk factor will become combined with other risk factors involved in the political governance of the matter, in the global evaluation. Saying that a mathematical analysis enables us to attribute a risk percentage is, therefore, meaningless, irrespective of the measures taken, of the set of rules and in particular of the evaluation of human reliability in the matter. Chernobyl, which was earlier mentioned, was a human error, the Challenger shuttle accident, a series of human errors, the Fukushima business, also a story of human errors… The list can be indefinitely added to. But we cannot hide behind this excuse of the human issue because everything boils down to human questions, and the human risk factor will always be there. It has to be taken into account in our evaluations. But is it quantifiable? And how? All this can be argued, but indeed it is not part of the initial mathematical model.

Convergence Between Fundamental and Applied Mathematics

Sylvestre Huet: Is there some sort of convergence between the major mathematical needs in industry and in nonmathematical research and those that mathematicians themselves consider to be the major current problems in mathematics?

Cédric Villani: They overlap but they do not fully coincide; they are not totally separate nor are they mixed up. I am going to provide some examples. The climate is both a mathematical and a social problem. It is a global problem: it is a matter of understanding and integrating familiar models using partial differential equations.

Models coming from fluid mechanics are not always well understood. The resolution of the Navier–Stokes equations in its theoretical version has a prize attached to it,[28] but even the issue of turbulence and the statistical behavior of a fluid is still not understood—we still don't even know what the correct mathematical formulation of the problem is.

On the question of integration: understanding how a complex dynamical system is going to work, what the good concepts are, whether it is stable or not. These are questions that arise both in mathematics and in real life.

The most famous current mathematical problem, which we cannot dispense with, is the Riemann[29] hypothesis, which says that the zeros of the Riemann zeta function lie on a critical line of the complex plane—except for trivial zeros that are integers. Though this is quite a misrepresentation, it can be said that this problem concerns the distribution of prime numbers, whether their global, i.e. in some sense statistical, distribution is strictly speaking, completely random. A priori, this problem has no impact on daily life. Some years ago, a leading Anglo-Saxon daily newspaper had run a headline saying that the solution to this problem could lead to a major catastrophe on the internet because they had in mind the connection between prime numbers and cryptography; but this mostly amounts to journalistic delirium. However, this problem has consequence on our understanding of some phenomena in quantum mechanics, on our comprehension of energy levels of atoms and in particular the spacing between these levels, because there are connections between this problem in number theory and these problems in quantum physics…

Pierre Cartier: These connections are known to exist, but are not explained.

Cédric Villani: Indeed, it is one of the greatest mysteries of 20th century mathematical physics. The Riemann hypothesis is the most famous, the most exasperating as well because it has been around for a long time—since 1859; it is fairly simple to state, but no one has any idea where to start; there are ramifications in mathematics, a very clear connection with theoretical physics; it is a real challenge, one that mathematicians would sacrifice everything for!

Jean Dhombres: In 1900 Hilbert said that it was the next century's problem…

---

[28] The resolution of the Navier–Stokes equations is one of the seven problems stated by the Clay Mathematics Institute in 2000. A correct solution to any of the problems is endowed with one million American dollars.

[29] Bernhard Riemann (1826–1866), German mathematician whose work was essentially in the area of analysis and differential geometry. The famous hypothesis named after him has still not been proved and is one of the famous 23 problems set by Hilbert as well as one of the 7 millennium problems.

Cédric Villani: And today, the question arises whether it should not be pushed back by yet one more century!

Pierre Cartier: But on the other hand, this problem has inspired so many things... Being incapable of proving the Riemann hypothesis, we have invented avatars of his famous function, for which we have formulated analogs of the hypothesis. The first to have done this is Emil Artin in his 1921 thesis; a first wave of research, by German specialists of the time, gave very positive results that André Weil's work immediately surpassed (results announced in 1940 and definitively proved in 1948). The story gets a new impetus when Weil himself formulated a broad generalization of the problem. This spurred Alexander Grothendieck on to undertake the task of entirely reshaping "algebraic" geometry (introduction of coordinates and algebra into geometry: Descartes' geometry!). This research project was completed in 1974 with Grothedieck's and his student Pierre Deligne's definitive results. But reshaping geometry goes beyond the initial motivations, and the exploitation of its consequences is ongoing and far from finished.

Jean Dhombres: With the Riemann hypothesis, we are in the theory of knowledge. In fact, mathematicians have never tried to hide that it is an essential problem, which would lead to a lot of theoretical developments, but which would not solve practical problems.

Cédric Villani: The most famous problems in mathematics are not necessarily those that have the largest number of applications, but those that are emblematic of a fundamental theory. For instance, Fermat's[30] theorem is emblematic of the development of number theory and of some essential problems in number theory.

Pierre Cartier: Fermat's theorem itself is rather anecdotal.

Cédric Villani: The regularity question for Navier–Stokes solutions is itself also anecdotal: we don't really care whether a velocity field of a viscous incompressible fluid always remains smooth or whether it develops spontaneous singularities, i.e. discontinuities during its temporal evolution. But it is emblematic of a major problem, that of a qualitative understanding of fluid mechanics, a problem still largely opens. There are other problems like that: the phase transition problem, emblematic of statistical physics...

Sylvestre Huet: It is strange to hear you say about this aspect of fluid mechanics: "we don't understand anything, it is still a largely open problem", when daily in the engineering departments of Boeing, EADS and every other aircraft manufacturer, planes are built using Navier–Stokes equations in computer simulations. On what grounds do you justify the terms "we don't understand anything" when the current

---

[30]Fermat's theorem, stated in 1641, was only proved in 1995 by the British mathematician Andrew Wiles.

solution, which you clearly consider almost nonexistent, is already sufficient to design an airplane?

Cédric Villani: I don't consider it as "non-existent"… An incredible amount of imagination, of sophistication and of fundamental development in numerical analysis has been marshalled to get where we are today. But this has been achieved by building on a number of presuppositions that are by no means a foregone conclusion, on situations that are not understood. However, if we understood them, we could perhaps go much further. So of course, we construct airplanes, we know how they behave in a given environment, under turbulent conditions. But it is first necessary to understand what turbulent conditions are, how to summarize turbulent conditions other than by computer simulation. The availability of "recipes" based on observation does not exclude the fact that we still have not understood the reason why it happens.

Gerhard Heinzmann: Now it is no longer the mathematician speaking but the philosopher! Indeed, statistical mathematics can be done without understanding its foundation. The engineer's mathematics is not necessarily the same as that of the mathematician…

Cédric Villani: … but on the one hand they converge. Once the reason why a theory is valid has been understood, it is possible to know the scope of its applications and when it will cease to be valid. In financial mathematics, people have performed ghastly calculations based on computable theories, on Gaussian and other statistics, until it was realized that when statistics are not Gaussian, everything falls apart; when it was realized that events supposed to occur every hundred years could happen several times in the same day, which clearly showed that the model did not hold water. And behind, there was a theoretical problem in understanding the limits of the paradigm being used. More than twenty years ago Mandelbrot had already understood that stock market phenomena implied a special emphasis on rare events, and that a Gaussian model was not adapted! Let us return to the example from aeronautics: it is well known how to construct the wings of an airplane with a given velocity field, but how can we tell how the airplane is going to behave in a different context? Turbulence engineers face this issue: "We have this ghastly turbulent flow and we do not know what is going to happen: how can this data be summarized to provide a code modeling another part of the aircraft?" It is a real problem. Engineers will disagree on the answer. There are no simplifying rules.

Jean Dhombres: Is this not where the difference of standards lies, for knowledge, but also for safe use, between a model and a theory? For Galileo, the time-squared law of falling bodies is not a model, but a genuine law… provided of being in a vacuum. In short, its safety was verified by space travel. Cédric essentially said that only models of fluid resistance are available.

Sylvestre Huet: What are the other industrial and technological challenges requiring innovative mathematical tools nowadays?

Cédric Villani: At present, among major technological challenges on a global scale, there is—it has already been mentioned—the climate issue, but also the energy issue and that of the distribution of resources. In the case of energy, there is in particular the controversy surrounding thermonuclear fusion: how to achieve fusion by releasing more energy that used for its realization? For example, take the project by inertial confinement. I visited the Lawrence Livermore National Laboratory (University of California, United States). In this huge building—having the dimension of three football fields—a battery of amplified laser beams are focused on small balls consisting of a mixture of deuterium–tritium of one millimeter in diameter, the aim being to exert a strong enough pressure on these balls for the fusion of the isotopes to take place. It is remarkable! Among limiting factors, physicists told me that at present their most serious worry was related to turbulence issues of instability type in fluid mechanics, because these small balls are surrounded by some sort of plastic layer which, once compressed, becomes fluid and forms convolutions which make it extremely hard to maintain an approximately spherical form. And added to this fluid instability, difficult to estimate, cracks occur in the solid structure of the small ball, which raises problems in the physics of continuous media. They thought they were facing a problem of fluid dynamics, of instability. They don't understand what would be the best course of action, the numerical code crashes because they are in extreme conditions that have not been calibrated. Under such conditions, they are systematically unable to obtain reliable simulations before experiments... So what should be done? No one knows. In all likelihood, the solution to the problem will be connected to theoretical progress, and partly to technical progress, through some fiddling which at some point will be more clever than at some other, and all this pertains to the whole.

Like in this example, mathematical problems often form an integral part of entirely contemporary issues, assuredly an extremely small part. In some sense they are like an emblematic island of an entire field in which one hopes that science is going to progress globally, in other words that this progress will englobe "fiddling", simulation, physical understanding and mathematical understanding.

Mathematics and Industry: A Culture Shock

Sylvestre Huet: Is there anything special in the organization of mathematical research regarding interactions with both private and public companies? There is a strong demand for mathematics in economic and administrative activities, which does not fall under public research. In your opinion, are these interactions operational? Between 1960 and 1970, in France major projects have been implemented, for example involving the SNCF and the CNRS on the problems raised by the TGV (high speed train), or agreements with the aeronautics industry, etc. The outcome has been clearly positive. Now, innovation is supposed to arise mainly in small and medium sized firms, but the structure of the relationship between public laboratories and a network of small and medium sized companies cannot be contrived in the same as agreements with large public firms. Do you think everything is fine or that things should be dealt with differently?

Pierre Cartier: The mathematical profession has become diversified with the emergence of "mathematical engineers"—to simplify things. Mathematicians now fall into two categories: academic ones and those who work in operational research with an engineering aim, for example in CNES (*Centre national d'études spaciales*) or in other organizations of this type. The latter have a perfectly sound mathematical background, but tend to leave the mathematical community.

Cédric Villani: Indeed, this is a matter of linkup. It is more or less clear to me that necessary action has been taken in the framework of the national loan scheme—and in mathematics it is perhaps the most tangible outcome—, I mean the creation of the *Agence pour les mathématiques en interaction avec l'entreprise et la société*[31] (Amies). This agency is supposed to maintain an information network, a meeting place between industrialists and mathematicians following the example of agencies abroad that facilitate the meeting of industry and mathematics. The purpose is to maintain the link through networking, and both topic-based and regional distribution by organizing seminars, meetings, working discussions, a bit like seminars: industrialists come to explain their problems to a mathematical public, mathematicians give talks in companies without any objective set beforehand, just to facilitate contacts. I think this is an interesting creation which is starting to operate—let's say we are attempting to catch up.

Another thing also important and difficult to quantify concern matters of discourse, image and culture. The historical gap between academics on the one hand and industrialists on the other must be bridged up in every possible way. In crude terms, let us say that the academic tradition considers industrialists as "blood-thirsty bosses", exploiters, people who neither respect intellectual property nor labor or scientists' rights. On the other hand, the industrial world regards scientists as ineffectual dreamers, part-hippy, incapable of adapting to present-day realities, quiet conservatives who do not wish to open up to the outside world. This description may be very crude, but the distrust between these two communities is undeniable, and besides, in my view, it is worse in France than elsewhere. It is a matter of culture. It develops over decades and depends on how the discourse is conducted and how it evolves.

It seems to me that to improve the image that each side has of the other, it suffices to make them meet regularly over a meal. Discussing over coffee or a drink will improve their mutual perception with all that this then entails in terms of collaborative projects. I think people are now aware of this need.

Moreover, the last joint initiative of the ministry of Culture and that of Higher Education and Research created a *Conseil national de culture scientifique, technique et industrielle* (March 2012). Separating once again scientific culture from general culture may be criticized. However, the positive side of a structure that does not separate science, industry and technique should be acknowledged. It is important

---

[31] Agency for the interaction of mathematics with business and society. Translator's note.

that these form a whole in people's minds—a whole does not mean that the same individual is going to have all the roles, but that each can identify with this or that. It seems to me the discourse is presently going in the right direction. Personally, I am proud to have been chosen by the *École des Arts et Métiers*[32] to write the preface to their commemorative book which appeared this year,[33] their aim is precisely to show the alliance between researchers and engineers both symbolically and historically speaking. Concerning the issue of discourse, this is intangible, but important.

I am more perplexed about institutional matters concerning the mechanisms needed for agreements between industries and others. For example, many people are convinced that the Ph.d. not being officially recognized by industry is a major drawback, but I do hold have a clear-cut view on this issue. In fact I used to think this should be the case until I came across the hearing organized by the *Office par-lementaire d'évaluation des choix scientifiques et technologiques*[34] during which representatives from doctoral students' associations said that the problem was not institutional, but cultural. And that disciplines in which this degree is the most recog-nized do not coincide with those in which the integration of doctoral students takes place in the best possible way. They added that half of the doctoral students are recruited for companies as executives on the basis of skills different from those they developed during their thesis, and they usually perform well.

Sylvestre Huet: The issue of the recognition of the Ph.d. degree through collective or company-level agreements is not really about hiring, but about the level at which the hiring takes place. I talked about this recently with the head of *Telecom ParisTech*: I asked him how many of his engineering student study for a Ph.d. The percentage he gave me is ridiculously low, something like 12 %. I asked him why? He replied: "Its very simple, when I introduce my engineer-students to industry bosses who are going to recruit them, they are told they can study for a Ph.d. if they wish to, but they won't be recruited with a higher salary than those with solely an engineering degree, or only marginally so." In these circumstances, why should the young get a Ph.d? The lack of practices that would prompt at least the best to decide to study for a Ph.d. is regrettable. What could justify a Ph.d. in the eyes of an engineer? Intellectual satisfaction and recognition by industry that would recruit him with a higher salary, justified because the student who slogs over his thesis during three years actually works far more than a young engineer during his first three years, and because a doctoral student has learnt what research consists in, how the academia functions… Thus, when the doctor–engineer enters industry later and has assimilated the relationship academia–industry, he becomes a choice asset. Everyone knows that the main difference between France and Germany is that in the latter engineers overwhelmingly hold Ph.ds. Where can we find here this recognition of doctorates

---

[32]School of Arts and Crafts.

[33]Olivier Vercherand, Anne Téqui, *Arts et Métiers—L'école de la technologie*, Le Cherche Midi, 2012.

[34]Parliamentary Office for the Evaluation of Scientific and Technological Choices. Translator's note.

in the sciences by the private sector or even by administration? Very few companies take this into account in their pay scale.

Pierre Cartier: Its true that bridges are not easy to set up. About ten years ago, I would gladly have spent one or two years in industry, but it was all terribly complicated! I kept repeating that I was truly interested in working in industry for a while, but everyone laughed...

Cédric Villani: I'm not surprised since general opinion regards a doctor going to industry as a failure—and usually this is also the manager's opinion.

Sylvestre Huet: If even only 50 % of student-engineers did a thesis, it would be normal after having finished a thesis to be able to go to industry as a researcher or an executive, and subsequently becoming a manager. If this is how things were, academia would be considered one possible path after a Ph.d., but not the only path.

Pierre Cartier: Unfortunately, *Polytechnique* sets a bad example. Not only do they literally despise Ph.ds, but the school has recourse to a form of corruption—and I have been in *Polytechnique* for sufficiently long to know. Take the example of my grand-nephew who studied at *Polytechnique* and who, after having been admitted, came to see me and asked: " How much do you earn?" That day I thought to myself: "Its hopeless!" I recently saw him again. He is a manager at Sanofi, and when I asked him why he had not studied at *HEC*[35] or *Sciences-Po*[36] or *Sup de Co*[37] as they would have been better adapted, he laughingly replied: "There are only Polytechnicians among Sanofi managers. If I was from *Sup de Co* or *HEC*, I would have remained at a lower echelon." The situation is the same in *SNCF*,[38] the "polytechnic mafia" holds all the key positions. *Polytechnique* students I managed to attract to research were considered by their classmates as gentle dreamers. And anyhow, in terms of career, this choice entails sacrifices, especially financial ones: those who join the ranks of academia earn half as much as those who join a company.

Jean Dhombres: There is a genuine institutional problem, in particular on the length of studies in engineering schools. At the *École polytechnique*, it is two years, followed by two more years in an application school. That is when a thesis could be done if application schools played by academic rules. But they are unable to as they depend on professional bodies. In my view, it is a typically French problem, an organizational problem "à la française". How have these professional bodies been formed until today? They have been formed to restrict competition, and hence quarrels, between the different bodies. Hence each of them has been given some space to do as they please and which doesn't encroach on that of another body: the *école des Ponts et*

---

[35] *École des hautes études de commerce* (an advanced business school). Translator's note.

[36] *Institut d'études politiques de Paris*. Translator's note.

[37] *École supérieure de commerce*. Translator's note.

[38] *Société nationale des chemins de fer* (National rail company).

Chaussées has the right to produce art works… but not to deal with ships, etc., which is not stupid for sound management. The problem is that if someone is gifted and interested then he should be able to go from one to the other… And professional bodies also have to learn to evolve according to technical evolution. Now, French mentality tends to consider all evolution as some sort of disturbance of the established order, and we often solve this basic issue by permanent agitation (reforms are no longer exceptions, but an existential rhetoric).

Gerhard Heinzmann: All this seems somewhat *"Ancien Régime"*, ill-suited to globalization…

Mathematics, an Ancillary Subject to the Other Sciences?

Sylvestre Huet: If we now consider the research sphere, it becomes apparent that various fields are currently halted, for example physicists in their attempt to unify quantum mechanics and general relativity, or biologists in their attempt to exploit massive parallel sequencing. Do you feel they have a need for new mathematics or will these sciences find the solution within their own (experimental or observational) dynamics?

It may well be asked since past experience in this matter is rather ambiguous. Thus Einstein is behind tremendous progress in physics, though he only used relatively old mathematics to formulate his theory of general relativity.

Pierre Cartier: The mathematics used by Einstein dated from fifty years before, but it was still necessary to be familiar with it and to go and look for it on the correct shelf!

Cédric Villani: Ricci curvature[39] was still more recent than Riemann's articles, from about 10 to 15 years before Einstein's theory was published.

Jean Dhombres: It was mentioned earlier that differential and integral calculus arose from the need to explain the Solar System based on equations from dynamics. But when you analyze the facts, you find that Newton had already acquired the basis necessary for differential and integral calculus before starting anything in astronomy. But he did not publish his results as he simply did not find them interesting. It is only afterwards that he discovered the importance of differential and integral calculus, when he realized it could be applied "to the real world". In other words, he came to trust his differential and integral calculus once he had made it live in a major astronomy problem.

---

[39]Ricci curvature: a notion of curvature derived by Ricci-Curbastro from that of Riemann, and which is the geometric cornerstone of general relativity.

Pierre Cartier: The aim was to recover Kepler's laws the latter had formulated based on astronomical observations.

Jean Dhombres: Precisely! And this example, so venerated by positivists (and which has disappeared from secondary school syllabi though it has been part of the claim to modernity and secularity), shows that invention is not necessarily motivated by a need; understanding the invention may only occur when its usefulness is realized.

Cédric Villani: Let us take a more recent example, that of wavelets. It is a beautiful story, with motivation coming from the industrial sector since the aim is to find new methods for the exploration of oil fields. In the early 1980s, Jean Morlet,[40] at the time an engineer with Elf Aquitaine, proposed the following idea: instead of merely disturbing the soil in a violent way and analyzing the reflected waves, why not send a frequency modulated wave train and see how they are transformed, with the idea that fine-tuning the dynamics of time/frequency will lead to a more refined and efficient analysis than merely sending a deflagration.

Pierre Cartier: In fact, Gabor[41] came up with this beautiful idea of wavelets thirty years earlier, but the promise this idea held was indeed not realized before later with Morlet, then Meyer.[42]

Cédric Villani: That's correct. A coevolution of industrial motivations, mathematical development, repercussions in mathematics and industry then followed. As a result, today they are found in some standard coding techniques…

Pierre Cartier: And in linguistics: phonological analysis of French and now of other languages is carried out using wavelet methods. Today, the phonemes that were earlier taught in linguistics are represented in a two-dimensional table in which 400 clusters describe the basic phonemes in French—including their variations. And these variations have been so well analyzed that the most efficient and the most reliable security systems are the ones based on voice recognition.

Cédric Villani: At a biological level as well. As Meyer said: "with wavelets, we started to reproduce what the bat's brain does": the latter uses wavelength and frequency modulation to detect its prey since to do so it needs to analyze its location as well as its speed and work against the uncertainty principle.[43]

---

[40] Jean Morlet (1931–2007), French geophysicist who was a pioneer in the field of wavelet analysis (equations used in signal processing).

[41] Dennis Gabor (1900–1979), Hungarian physicist who received the Nobel Prize in physics in 1971 for his invention and development of the holographic method.

[42] Yves Meyer, French mathematician born in 1939, a specialist of wavelet theory.

[43] The uncertainty principle (indetermination principle), formulated in 1927 by the German physicist Werner Heisenberg, states that it is not possible to simultaneously measure the position and momentum of a particle with absolute precision. The proof of this principle is fully mathematical,

Sylvestre Huet: The question of what mathematics has to offer is implicit in all these interactions between mathematics and other disciplines. How can we make biologists, physicists, linguists, economists, etc., aware of everything mathematics has to offer and ensure they do not miss a solution? How can this interaction be organized? Should mathematicians be spread among other faculties? And is this possible to do given the current structure of scientific organization? What are your views on this matter?

Cédric Villani: Indeed this is not a trivial matter, and it does not only arise in science. The first problem faced by someone setting up his business is to gather together various skills and to make specialists who are not acquainted with each other's work communicate mutually—the law specialist does not understand anything about accountancy, the accountancy specialist does not understand anything about management and so on. And indeed, the same problem has to be resolved in science: how to ensure that mathematicians communicate with physicists, that physicists communicate with biologists and so on.

Pierre Cartier: This is a ferryman's role.

Cédric Villani: Yes and you have used a good term. People, not ideas, need to be transported because ideas are carried by people and hence the exchange happens during discussions. People need to be put in physical contact and this is not easy. It is necessary to communicate, meet, give people the opportunity to meet, organize general seminars, public conferences, install coffee machines in meeting places…

Sylvestre Huet: Is this sufficient or is it necessary to do more? For example, might mathematicians be recruited by geology or linguistics laboratories as mathematicians, like industry does?

Gerhard Heinzmann: The Henri Poincaré Archives in Nancy has been recruiting mathematicians, physicists, historians, sociologists, philosophers since 1992 and thereby it is now an interdisciplinary place characterized by the dialog between specialists. But there is a need to go further. To be consistent, the traditional structures of a "Faculty", of a "scientific cluster" or of a "subject area" should sometimes be replaced by a structure based on a "Centre" which would allow to fight against fragmentation and which would have a permanent structuring effect. Within the human and social sciences (HSS), this was done with the "Houses" of the Science of Man that are vital interdisciplinary centers in HSS. Abroad, "inter-faculties" have been developed, where both cultures, HSS and "hard sciences", are part of the interdisciplinary approach.

(Footnote 43 continued)
using Fourier analysis, but this first requires a mathematical formulation of the physical concepts of position and momentum.

In my view, as far as possible, a topic-based interdisciplinary approach should be organized, as an integral part of the scientific process, especially in cutting-edge research that requires an organized interdisciplinary exchange. Some students stand out by their smartness and their attitude towards the "two" cultures. From the beginning of their studies, they ought to be given the opportunity to consolidate their independence of mind and their disregard for dogmatism. They should be given the possibility to participate in basic teaching, based on research done in the interfaculty whose council should be an advisory body of the president of the university. There are some twelve initiatives of this type in Harvard related to the environment, ethics, health, human rights, microbiology, neuroscience, etc. I hope that the legal framework will soon allow some French establishments to set up such interfaculties and I hope this for the University of Lorraine.

Pierre Cartier: I would say that this is the problem of scientific monitoring. All laboratories and large institutions should have a scientific monitoring body to detect what is happening—and not only in mathematics within other fields, but within its own field—because fields are very fragmented and it is necessary to be able to climb over fences. Scientific monitoring is a delicate problem and not new.

Jean Dhombres: This problem is indeed far from being new. Besides, the rules of the French Academy of sciences, laid down in 1699, does not require Academicians to publish articles, but to make known new publications, etc. that have caught their attention.

Cédric Villani: The seminar Bourbaki follows the same idea: each speaker presents something he has found interesting.

Jean Dhombres: We are accustomed to speaking about the "scientific community", but it is not so simple… In Galileo's days for example, the "scientific community" had no consistency, and at best one can name it a "Republic of Letters". It may be that because of Galileo's conviction in 1633, scholars became aware that they ought to organize a security cordon. But it was not until the end of the 17th century and Descartes'[44] influence in France and that of Bacon[45] in England, that one hears of the necessity of a place—Descartes talks of a society, Bacon of a chamber—where what is new can be discussed and estimated, to eliminate elements that do not meet the requirements of scientific legitimacy, which were unclear at the time and are still even more so in this day and age. Today, the scientific world in all its complexity, I want to talk of the men and women it consists of, is facing a problem of quantity, of number, hence issues of connection, network, and also of power…

---

[44]René Descartes (1596–1650), French philosopher, considered to be the founder of modern philosophy.

[45]Francis Bacon (1561–1626), English statesman and philosopher, one of the pioneers of modern scientific thinking.

Pierre Cartier: In developed countries, researchers represent about 0.7 % of the labor force.

Jean Dhombres: This is not a negligible figure. How many farmers are there nowadays?

Sylvestre Huet: Farmers represent around 3 % of the labor force, so about four times more.

Jean Dhombres: Still these are comparable figures. And this figure gives rise to lots of management issues.

Cédric Villani: Managing both this large number of researchers and research is the duty of the State. How is a different matter and this is why a genuine research policy seems essential.

Scientific Policy: Organization and Development of Mathematics Research

Some Organizational Principles of Mathematics Research

Sylvestre Huet: We have provided many examples for the idea of the extensive use of mathematics in scientific and technical activities in our societies, and the distinction between fundamental research and the use of mathematics is clearly not a distinction between people, but an operational distinction between activities, which is not the same thing.

Cédric Villani: They are the two extremes on a continuum. And I think that in some cases concepts are going to change according to the operational aspect for example. Similarly, the fact of teaching is not neutral: it has consequences on future and ongoing research, on our choices—what is important; what is less so? It is not all that clear-cut. At my level, synthesis work accomplished by others greatly influences my personal production.

Sylvestre Huet: This brings me to the question of whether you consider France's scientific policy to be good or bad? This question has been raised several times in the last thirty years by scholarly societies in mathematics. According to you, is what needs to be done being done? And if not, what should be done in the various areas of training, research and mathematical applications?

Cédric Villani: My position on these questions has always been rather qualified, and before giving the floor to Pierre whose position will surely be more radical, I would like to recall some general principles about mathematical organization: first, it is important to keep research units in mathematics, in other words, places where mathematics is done, places where different branches of mathematics communicate, in short crossroads.

Pierre Cartier: I fully agree. Overspecialization of mathematicians should be avoided.

Cédric Villani: This defect became common in the 1970s when combinatorics laboratories, analysis laboratories, laboratories of this and that were set up.

Pierre Cartier: And it is still unfortunately the case nowadays.

Cédric Villani: I am perhaps influenced by my experience in Lyon, which is at the other extreme, with a more comprehensive approach: we have a mathematics department, many common seminars and interactions between groups. It was a huge culture shift for me when I went from the *École normale supérieure* in Paris to the one in Lyon; I went from a situation where we were separated into corridors according to our discipline, to a situation where we were all intermixed.

Pierre Cartier: I can relate my recent experience. I am just back from three months at the University of Gottingen and to my dismay, I found there many small groups whose work is certainly very good, but separated. As someone external, I did not really manage to integrate because there was no overall life. And a scientific center where there is no communal life is a disaster. I have always been in establishments where the various branches of mathematics were not separated, nor was mathematics kept apart from other departments (physicists, chemists, etc.), whether at the department of mathematics and computer science of the University of Strasbourg, or at the department of mathematics and theoretical physics of *Polytechnique*, or at the *Institut des hautes études scientifiques* (IHES). I have lunch every day with my friend Thibault Damour, who is an astrophysicist, and we do not lack discussion topics; I attend his talks, he comes to mine… For me, the most important thing that Russians have given us is this basic example of interaction.

Cédric Villani: I entirely agree, and this is the second point I would like to insist on—and in this respect too, Russians may have pushed this point to its limit—, namely the notion that a department revolves around a seminar. It is an opportunity to get together, a social moment which proves to be much more structuring for a mathematics laboratory than a scientific collaboration. It is somewhat of a mathematical specificity—and possibly also a theoretical physics one. It is the highlight of the week when we all meet. In the Russian school, the seminar aspect has been exploited to the full, it was even its favorite means for transmitting information (more so than through publications). It is a social event where people interact, and that can occasionally be quite violent; the speaker may be vehemently attacked, quarrels may arise between various parties… This form of direct exchanges can be very efficient.

Pierre Cartier: I would add that for the Russians, external social constraints of Soviet society were such that seminars acted as an outlet for frustrations.

Cédric Villani: The third basic organizational principle—which has been followed in France fairly well for a long time—is to not cut research from teaching: the "normal" career of a mathematics researcher is to be a university professor and not a full time researcher. I even belong to the extremist school which considers that CNRS research director positions should be abolished—or their numbers substantially reduced—and instead teaching hours should be reduced in universities. I myself hold a university position with no teaching, which does not prevent me from teaching, but my basic title is university professor, and I prefer this title much more than that of a researcher (from a statutory point of view) who would also give courses. I am a teacher-researcher and in my view teaching is so fundamental that I want this to be written in my contract.

Jean Dhombres: As for me, I only accepted a CNRS research director position after being elected to a position at the *École des Hautes études en sciences sociales* requiring me to teach, though in the form of seminars.

Sylvestre Huet: Is it possible to give an idea of the number of mathematicians working in CNRS and in universities? Do you find current scientific policies satisfactory (or not), in terms of employment policies for instance?

Pierre Cartier: I would say that, very roughly, there are 500 CNRS mathematicians and about 3000 university ones.[46]

Jean Dhombres: Just to place the question in its context, I remember that in 1970, in Paris, there was a small office at the IHP[47] housing about the one hundred CNRS researchers in mathematics in Paris, of which I was one.

Cédric Villani: Another mathematical specificity is that CNRS is the only national operator, which obviously gives it significant leverage. It is a "brand", a respected institution, and for a very long time, CNRS conducted mathematical policy at a national level in a fairly independent manner, without necessarily taking into account what was being decided for scientific policy as a whole. In particular, the number of research directorships for mathematics professors was very low. In any case this is the subject of the ongoing debate: over the last two years, the number of research directorships in mathematics has increased. Many, including myself, think that this number should have been kept very low, entry level research positions should have been increased and senior level ones decreased. This point does not enjoy unanimity.

---

[46]According to the *Institut national des sciences mathématiques et de leurs interactions* (INSMI/CNRS), there are 400 CNRS researchers in mathematics, about hundred postdocs, and more than 1400 doctoral students. Researcher-mathematicians within universities and other establishments amount to 2750 (figures from 2011).

[47]Henri Poincaré Institute.

Gerhard Heinzmann: It would be interesting to know if in mathematics there is a difference in the quality of the average productivity of a CNRS researcher and that of a teacher-researcher. Do most mathematicians who matter in France hold a CNRS position?

Pierre Cartier: And what do you think about the development or not of the *Institut universitaire de France* (IUF)?[48]

Cédric Villani: In my view, the number of members should be increased, and the process of democratization should be reinforced. I think this was a huge, remarkable progress.

Pierre Cartier: And one of Claude Allègre's[49] rare good achievements.

Cédric Villani: The *Institut universitaire de France* is truly a model, which does not severe contacts with the university and especially with students, in other words with "real life". For real life is the students who will become the main body of scientists, not only within research. This contact must be maintained. With the IUF, teaching is maintained, though hours are reduced, and allows itself considerable leeway, while sparing itself the psychological problems often encountered by full time researchers, who sometimes have problems due to the lack of constraints, as being totally free is terrifying, and often destabilizing.

Pierre Cartier: There is also a financial aspect which is never mentioned, but when a CNRS researcher gives a class, he gets overtime pay—which does not always make colleagues happy.

Sylvestre Huet: Are flows (hirings and terminations) better managed today?

Pierre Cartier: No, it is quite disastrous…

Cédric Villani: Indeed, it is not very well managed. For example, if you are the head of a mathematics laboratory, when x number of people retire—and x may be large, for in big departments it happens that eight positions need to be filled the same year, which is a huge number for a laboratory. What should be done? The right thing would be not to fill all positions if there are insufficiently many candidates or they are not satisfactory. But the forces faced by universities make it very hard to say: "Please freeze this position, let us wait till next year for better candidates." So usually it ends

---

[48]The purpose of *Institut universitaire de France* (IUF) is to facilitate the development of high-quality research within universities and to enforce an interdisciplinary approach. It was created by the 26 August 1991 decree as a department within the ministry responsible for higher education. Its members are all teacher-researchers who continue to teach in their respective universities, but with an adapted timetable, and receive funding for specific research projects.

[49]In 1991, Claude Allègre, an internationally renowned French geochemist, was special adviser to Lionel Jospin, then Education minister (*Éducation nationale, de la Jeunesse et des Sports*).

with: "I am too afraid that this position will be closed, so I am going to fill it, even if it is not in our best interest." This situation can prove to be very serious because once you have recruited the wrong person, he is going to be there for a long time!

Jean Dhombres: Its true that it is almost impossible to freeze a position and wait for the desired candidate until the following year because of preset budgetary reasons and yearly allocations, etc. Luckily, within such an administrative structure, we still manage to "fraction" out, in other words to transform a professor's months into months for invited academics…, but to do this you need to be trusted by your colleagues.

Pierre Cartier: These situations sometimes arise at the *Collège de France*, where the professorial chairs have no permanent titles. Each time a chair becomes vacant, they have to vote on the title it should be given. And from what I know, it can happen that an academic discipline gives a chair to another with future reciprocity.

Jean Dhombres: This is more or less how the *École des Hautes études en sciences sociales* (EHESS) functions, since it is the candidate who defines the chair profile he will hold and so is elected on this project.

Pierre Cartier: But this is only possible because they are small institutions, and are not really representative.

Cédric Villani: Precisely, and that brings us back to the issues I raised earlier on the amount of leeway high schools have. There is also another side to it: the problem of flow management according to age. For a long time, there was little pressure on senior teaching positions, there were many vacancies and not many candidates. But when large numbers do not find any positions, a whole age group can end up "on the street", that is to say without a job. It is important to be able to smooth out this problem over time. In fact this problem is encountered in economy: how can it be explained to people that it is necessary to save a slice for bad times? I would say that this is a matter of governance. In mathematics, we have not managed this flow very well. I don't know if other disciplines have done better. A matter of mutual trust between disciplines, with university presidents too, is involved. It is perhaps one of the main problems we have been suffering from.

Pierre Cartier: Politicians did not help us either, given their short-term view related to election timetables. I remember a meeting with Jacques Valade when he was minister of Higher Education and Research, in 1987–1988: Several of us, researchers and academics, were explaining what we would need "5 years, 10 years hence"; he replied: "But who will be minister in six months?"

Cédric Villani: This brings us back to the issue of trust, of governance, and of who embodies them? In my view, one of the pillars of the system is the university president, who embodies the trust of the various departments, and also arbitrations. It is a basic moral role. You need to be both strong and open-minded. I was the president of the Board of specialists in mathematics at the *École normale supérieure* of Lyon for six years. It is a very small institution where we were very sensitive to recruitment issues over time, which led us to handle things completely outside the administrative framework—even more so at the University of Lyon—to get out of situations where, for example, we did not have a position for an excellent candidate who had also been offered a position elsewhere; the Board met informally and people agreed between themselves that the following year's position would be assigned to this candidate. So a promise was made to the candidate: "Do not accept the position you have been offered elsewhere, next year this position will be yours, we promise you." Acting in this way was outside all norms, all laws, but we had assumed all responsibilities in this matter. This is also what politics consists of. In the end, everyone wins—if it goes well, if there are no "traitors", because someone could have sought legal recourse… And having to respond to a situation in a hurry does not exempt us from having a long-term vision, even if this means adapting to circumstances as one goes along. This naturally leads to the issue of how much planning is desirable, how much should be done, and how much can be achieved? This is where one often comes across completely absurd situations, for example when people are asked to anticipate with a wealth of supernatural details what they are going to do in their research project.

Gerhard Heinzmann: I totally agree with Cédric and, on an even smaller sale, I pursued the same policies in Nancy: you must not wait for applications to a position, but attract candidates and say this loud and clear so that everyone, even the students, can form an opinion and argue during preliminary and informal meetings.

Sylvestre Huet: In fact what are your views on the ANR (*"Agence nationale de la recherche*) projects which seem to completely run counter to what should be done, both because they are all short-term ones, and because they greatly increase the workload due to the need to present projects when the vast majority of them are going to be turned down. Is it not a huge waste of time and energy?

Cédric Villani: Thats true, but at the same time I think that setting up the ANR in itself was a good thing. In discussions with the management board of the ANR, you come across such a spirit about development possibilities, still fresh, with an open-mindedness typical of new institutions, and which you do not find in CNRS. It is far easier to get things going at ANR level than at CNRS level because tradition has not yet set in. The ANR is still trying to find its path but I found its recently nominated management remarkably open-minded—though an enlightened management is no guarantee for an enlightened institution as a whole. And finally I would like to add that when calls for projects *Laboratoires d'excellence* (Labex) were being set up, when we received the letter written by the ANR asking us to anticipate our budget for the next 10 years with a wealth of details I would have been incapable of providing for the ongoing year in my institute, we thought it was a joke. The ANR says that they did not write it themselves—the issue of a dilution of responsibility, of the causal pyramid, is always present. As for the issue of responsibility, it is not very clear, perhaps is it more subtle than it seems? Thats my "straddling the fence" side...

Jean Dhombres: Not being bound to silence, I will be even more harsh. Indeed, contrary to every practice I have come across until now, and even within the *Conseil national des universités* (CNU), I recently encountered explicit breaches of research ethics due to the competitive culture. Negative evaluations of a research group were given to support the research laboratory supposed to judge in all good faith; I even heard a laboratory being blamed for having too much access to funds, when the evaluation should have addressed the proper use of these funds. Peer evaluation, one of the oldest rules to maintain a balance within academia is being devalued in France.

Is Mathematical Research a Place of Freedom?

Sylvestre Huet: Can we now return to mathematical research: how it is done, how it is experienced at individual level, as regards motivations, rhythm, results, but also as regards needs and tools, in other words: how do you work?

Cédric Villani: There is no general method, styles depend on individual mathematicians, and each mathematician's style will vary depending on periods. Motivation can come from a project, an encounter, sometimes from an idea we want to delve into further, other times you think hard and you get headaches trying to find the right concept, sometimes you calculate until you get a spark of insight...

Sylvestre Huet: Can you give some personal examples of how you work?

Cédric Villani: In my case, what is striking is the importance of chance factors, of chance meetings with this or that person. I have often thought about this because of my history. I am going to start with the elaboration of my most cited article,[50] perhaps not the most important but the most cited. This article, written around 2000 together with a German collaborator, Felix Otto, developed a new approach to a certain class of functional inequalities. Put simply, they form a bridge between questions of a scientific nature and questions of an optimisation nature (optimal transport with respect to certain parameters related to geometric inequalities of isoperimetric type in a broad sense). What is interesting is to see how this paper came about. I met Felix Otto in 1999 in a workshop organized by my tutor, during which he presented a fairly new, interesting formalism, which grabbed my attention; and in addition, I was busy reading and sifting through Michel Ledoux' lecture notes (probabilist from Toulouse, well versed in analysis), who happens to be one of my mentors, and I came across this inequality in these notes. I was trying to understand it—it is not my main

---

[50]F. Otto and C. Villani, Generalization of a inequality by Talagrand, viewed as a consequence of the logarithmic Sobolev inequality. *J. Funct. Anal.* 173, 2 (2000), 361–400.

subject—, and suddenly, I thought to myself: "That's funny, this inequality curiously reminds me a bit what Felix talked to us about." And I quickly understood within an hour that by using the formalism set up by Otto, I could find a new approach, a new proof of this theorem in Ledoux's course. Things then had to be put on a formal basis. I went back to Otto, we wrote the paper together, we delved further, but everything came about from the fact that I had heard Otto speak at a seminar a month earlier. This article immediately roused interest, and was straight away accepted by the *Journal of Functional Analysis*. And then, the first coincidence occurred: some months later, another of my mentors, Eric Carlen[51] who was on my thesis jury, invited me to Atlanta to give talks on the Boltzmann equation[52]—one of the topics I had worked on. I showed them my work on Felix's new formalism. It appealed to Wilfrid Gambo (Carlen's colleague), a specialist of optimal transport, who then invited me to spend six months in Atlanta. Once there, I felt that as I had to give a course, I might as well give it on optimal transport since that was the subject I was starting to explore. Allow me a little aside concerning this general principle which we should have mentioned earlier: the best way to learn is to teach. This has been known for millennia. So I started to write down my course notes, I fumbled, and finally this course turned into my first book on optimal transport[53]—a book that sold quite well and filled a gap since there were no general books on this topic.

Some years later, in 2004, I went to Berkeley, a great department where I got a bit bored because there were not enough interactions, not enough seminars, when a guy came into my office; his name was John Lott and was an invited guest at the Mathematical Sciences Research Institute (MSRI)—the equivalent of the Henri Poincaré Institute in the United States—, whose job was to get things moving by making people meet each other: " I am John Lott, I am a specialist in Riemannian geometry and analysis, I've read your articles, your book, we have a great program: we are going to do synthetic geometry with optimal transport." And he started to explain non-Euclidean geometry, what curvature means, by drawing triangles and then he said "we are going to tackle the Ricci curvature [curvature involved for instance in general relativity] as we did sectional curvature, but this time based on optimal transport; for our axioms we are going to take the theorems you proved together with Otto and this is what we would like to show…" And he proceeded to explain Gromov's[54] geometry—I was a complete ignoramus on this subject—and it was a revelation. It was extraordinary. And though I was not a geometer, I started to work intensively in this area. I switched over directions entirely… All merely because I met this mathematician, whom I saw twice for half an hour. We decided to work together and after four or five years, we published an article, which proved

[51] Eric Carlen, Professor of Mathematics at Rutgers University, Piscataway, New Jersey, United States.

[52] The Boltzmann equation, formulated in 1872, is an equation describing the evolution of a gas.

[53] *Optimal Transport: Old and New*, Springer-Verlag Berlin and Heidelberg GmbH & Co. K, 2008.

[54] Misha Gromov, French mathematician of Russian origin born in 1934 who received the Abel prize in 2009, known for his important contributions to various geometric fields. He works at the *Institut des hautes études scientifiques* (IHES), Bures-sur-Yvette, France.

to be a hit! We developed a theory based on optimal transport to detect curvature. In other words, how a geometric concept can be formulated in terms of optimization and transport. A theory which takes stock of my former loves, entropy and statistical physics.

Pierre Cartier: And what was Perelman's[55] role in all this?

Cédric Villani: Perelman was implicit. Subsequently Lott used some of our work to rewrite Perelman's proof. Which further strengthens my admiration for Perelman who saw things… I have no idea how he saw them, retrospectively it is possible to reinterpret because of this and that, but he did not have this and that.

Enumerating the coincidences that enabled the existence of this article show how incredible it was. This is also what mathematical research and research in general is, a matter of arranging coincidences: they cannot be foreseen, but you can ensure that as many as possible happen.

The coincidence still continues: following this work with Lott, I was invited to Princeton in 2009, and my stay there coincided with a time when I badly needed to be alone to think about an incredibly difficult problem, which required intense thinking. Princeton gave me this space which I could not have found in my usual surroundings.

So you see, sometimes I need to meet people, at other times I need to be alone, sometimes I need to develop my interests, at others take them further in a certain direction. Research amounts to all this. There is no normal practice. In year n, you have no idea what you are going to do during year $n + 1$ and it is better this way, it opens greater perspectives.

Sylvestre Huet: And this entire series of coincidences was possible because of the intrinsic organization of academic life which enables scientists to meet during seminars and colloquiums. In other words, coincidences may not be foreseeable, but at least meeting and exchange places and times have to be scheduled.

Cédric Villani: Quite, governance is mainly about setting up institutions where unexpected events can take place. To this end, conditions need to be met. I was saying earlier that the vital place of a laboratory is the seminar, where you come across other peoples' ideas, where you learn and which, occasionally, provides insight, things suddenly fall into place. Perhaps only 1 % of the content of seminars will be relevant to your research, but it changes everything, it is often the missing link. Trying to reduce a priori by taking only the useful would not work.

---

[55] Grigori Perelman, Russian mathematician born in 1966, who was awarded various international prizes, including the Fields Medal in 2006, but he turned them all down. He, in particular, proved the Poincaré conjecture in 2002, one of the seven Millennium Prize problems, each of which has been endowed with 1 million American dollars by the Clay Mathematics Institute.

Sylvestre Huet: And what about you, Pierre Cartier, what is your way of working and has it evolved with time?

Pierre Cartier: As for me, I would say that my experience is quite different from that of Cédric. I am very curious and an impertinent chatterbox. Feynman[56] used to say:

> How do we do it? It is not complicated: at all times we are thinking about twenty topics, and each time something becomes clear, we store it in a box. And after a while, the box is full and can be emptied.

There are no universal recipes, each person has their own. Me, I am a frontiersman. I was born near Luxembourg, where no one knows whether he is Belgian, French, German or a native from Luxembourg, within a family, where no one knew whether we were really Jewish, Catholic or Protestant, etc. I worked in Strasbourg—the perfect frontier town—for many years and scientifically I have always been in frontier areas, which my Bourbaki colleagues were not always happy about. The 1960s were a time of a profound estrangement between mathematical physics and mathematics. I remember giving talks at the Bourbaki seminar on probability or mathematical physics, which did not go down well. For example, when I talked about Spitzer's identity (in relation to probability theory), Dieudonné met me at the end and said: "This is *freshman algebra*", in other words, basic algebra. That's true, but to reach this ultimate simple form, it needed men of genius like Spitzer.[57] The ultimate form of a theory is indeed to be easy to understand.

Jean Dhombres: It always boils down to the thriftiness of mathematics. And I particularly appreciate the example of Spitzer's identity chosen by Pierre, because it reminds me of a remark by Gian-Carlo Rota on a similar contempt encountered, this time in the United States, more precisely at the Massachusetts Institute of Technology (MIT) in Boston, when they embarked on a combinatorial study of this identity to understand its significance beyond probability.

Pierre Cartier: I new pertinently well that I could give talks on mathematical physics in the Bourbaki seminar because of my strong position within the group. Concessions were made for me...

Basically my recipe is to proceed as Feynman said: at all times, I am thinking about two or three mathematical topics and I store my ideas in a box—probability, combinatorics, etc. I am obviously somewhat exaggerating to make my point. In fact, I am simply at all times in a scientific and technological watchful state. I have always kept close links with physics, especially at the time when I made a modest contribution to the development of radio-astronomy in France (in the 1950s), or when

---

[56]Richard Feynman (1918–1988), American physicist who received the Nobel Prize in physics in 1965 for his work in quantum electrodynamics; he was one of the most influential physicists of the latter half of the 20th century.

[57]Frank Spitzer (1926–1992), American mathematician of Austrian origin whose work was mainly in probability theory, in particular in random walk theory.

I worked (around 1960) with military engineers on coding problems which I used to solve within ten minutes using the field with four elements (all that was needed was intercepting two quadrics in four-dimensional projective space over a field with four elements!)... It was so wonderful to be the only one with an overall view and to discover that fairly complex mathematics could be so easily applied.

So I would say that developing connections is the most beautiful thing I have accomplished: connections between probability and analysis, between combinatorics and geometry... and it just happened like that. One day, together with Bourbaki members, I had taken a night train to go to a meeting and all of a sudden, in the middle of the night, I woke up my neighbor (it was Serre) to tell him I had found something important! Anecdotally, far from sharing my enthusiasm, he said that if it was that important I would still remember it the next day...

Cédric Villani: I feel perfectly at home here. In my case too, things I am most proud of are connections, like connecting optimal transport and geometry, together with Lott. The first piece of work I am proud dates from when I was 23: I was visiting Pavia where I had met an Italian researcher, Giuseppe Toscani, who was trying to solve a problem called Cercignani's conjecture, which he gave me to do as he was busy and I had time. I quickly saw that it was impossible to prove this conjecture, but that calculations I had been forced to do were the key to progress. Finally we solved the problem and the solution contained a formula—which we would have been incapable of saying what it covered—which connected entropy production in the Boltzmann equation, entropy production in the equation of plasma physics and in the Fokker-Planck equation. Deriving one of the equations with respect to another led to the third. It was an unexpected connection between objects and "bang!" it gave the solution of the conjecture. During my thesis defense, Meyer told me: "There are connections in your thesis." Twenty years ago, this piece of work would have been made fun of because of the presence of this element which seemed to come out of the woodwork, and which we had introduced ourselves. My whole career is thus marked by a series of connections.

New Tools, New Practices

Sylvestre Huet: Your stories clearly show that to make coincidences happen, people need to be recruited, meetings organized, interactive sessions set up, that researchers should not be compartmentalized, etc. But what are the material tools needed in mathematical research: paper, pencil, computers, etc. In what way have computers changed research practices in mathematics?

Cédric Villani: Computer science has turned everything upside down in terms of transmission and access. Talks can now be listened to later, thousands of kilometers away; collaborators can work together at a distance; information can be obtained within a day via electronic mail; results are transmitted to researchers throughout the world before they even appear in print in scientific journals.

Pierre Cartier: Today writing an article with a colleague living 3000 km away has become very easy because of the internet.

Cédric Villani: My best papers were written with collaborators thousands of kilometers away from each other.

Pierre Cartier: Writing articles has been facilitated by softwares such as LaTex. Similarly, access to documentation has completely changed, not to mention the organization of colloquiums. Fifteen years ago, this required a lot of work: unceasing exchanges of handwritten letters... Today, a few emails, and its settled! The quality of communication has greatly increased collaboration outputs... But this is not special to mathematicians, everyone extensively uses this mode of communication.

Sylvestre Huet: Has it speeded up work?

Cédric Villani: I want to qualify my answer. It has significantly speeded up work: you ask a question and you receive the answer within the hour—if you know whom to ask. While preparing my course notes a few days ago I encountered a problem—I am teaching the Kolmogorov–Arnold–Moser theorem (KAM) this year—, I started pulling my hair out because Kolmogorov's term does not apply to the Solar System. I emailed Ghys and I got the answer an hour later: "You should read Arnold, Féjoz, Herman, and you will find your answer: you will see, the Solar System is degenerative..." It would have required a significant bibliographical search, discussions with people... And here, I got the answer within the hour! This is a radical change.

Now the qualified part: Concentration is a major challenge for a scientist who really wants to see where he is going. And in this respect, it is painful! I have become dependent on my computer because as director of the Poincaré Institute it is my duty to follow what happens in real time, on a timescale that is not that of research. Its disturbing.

Pierre Cartier: This is why I have chosen not to have any internet connection at home, to be left at peace on weekends.

Cédric Villani: This I cannot do, otherwise everything collapses, my institute collapses! At least, this is an irrational fear that takes hold of me. It is a handicap and a major worry. When you are teaching, and you see students looking at their phones every ten minutes... I feel that as well, we are assailed by information which has the same effect as a drug. As soon as we have the time, we look at our phones to see whether we have received a message... This is something we should be very cautious about. I'm not saying its a bad thing, but that you need to be disciplined, and you need to be aware that from a certain viewpoint it acts as a brake because the amount of information is constantly growing. How many emails do we receive from here and there? A problem which in time would have been automatically solved can in no time become a galactic problem with emails in every direction! And when the problem is slightly serious, it goes all the way up to the Ministry, with very little informational content, which perturbs and prevents proper functioning. Similarly, at present, almost 95 % of communications throughout the world are spams. Some sort of informational background noise has come into being and has become a bondage. If I want to manage the flow of emails I receive, I daily need about two or three hours! You could say that its is marvelous that we manage to do so many projects simultaneously, but you could also say that this represents time spent away from reflection and strategy: we let ourselves be tossed about, we are far less able to see the forest for the trees. So there is an ambivalence, which is characteristic of all technologies, and which does not only concern mathematics. In mathematics, we feel it a bit more perhaps in the sense that our work (our concentration, our reflection over time) is perturbed, but this problem affects everybody.

Sylvestre Huet: A few years ago, I wrote a piece on Jean-Pierre Serre in *Libération*, where he says that ideas came to him while washing-up because it is a repetitive action which is not very intellectually demanding, or while watching advertising in a cinema hall. In other words, he is like Newton, "he thinks all the time".

Jean Dhombres: Or like Lagrange[58] who used to go to concerts only to be free to think about other things.

Sylvestre Huet: Besides, our last Nobel Laureate in physics, Albert Fert (one of the applications of his work can be found in mobile phone memories), was telling me that before receiving the Nobel Prize, he had always refused to have a mobile phone to be left at peace, in particular when traveling by car between Paris and Saclay, and that the more there were traffic jams, the happier he was since he could think for a long time without being disturbed.

Cédric Villani: Me, I always refused to have a mobile phone until I became director of the Poincaré Institute. I lost this determination then. No way out of it. The two groups of people who force you to have a mobile are—by experience—politicians and journalists, who need to be in contact on a time scale even less than the time needed to make administrative decisions!

Sylvestre Huet: In other words, one of the roles of institute directors, university presidents, and even politicians who are in charge of all this, is to ensure that the organization of scientific life allows researchers more time for reflection.

Pierre Cartier: I would like to insist on the fact that mathematics are acquiring an experimental component. First, we now have extremely efficient data bases—everything that used to be in catalogs or tables has been digitalized, and this has led to an amazing expansion of our capacity to store information. Let me give the example of a classics of numerical analysis called *Abramovitz and Stegun* (after

---

[58] Joseph Louis Lagrange (1736–1813), Italian mathematician and astronomer, who lived in France from 1788 onward, and whose exceptional work comprised every area of analysis and mechanics.

the authors' names); its a thick volume with numerous formulas, tables, etc. In the recently published new edition, numerical tables are replaced by an accompanying DVD and which only needs to be installed on a computer. Thus, more than half of the book has disappeared... but the thickness of the volume has increased! In fifty years, there are twice as many formulas and there are totally new chapters: random matrices, q-calculations, etc. And besides, a colleague recently presented us the online and fully interactive version of this book. This means that formulas can be created on request. Which has the advantage of getting rid of printing mistakes, of being able to zoom in, of getting a 2- or 3-dimensional graph, of computing numerical values, etc.

I did some experimental mathematics 35 years ago, I might as well say during "prehistory"... You then had to move about with huge cardboard boxes with perforated cards. Today, for Riemann's hypothesis, we get 60 million zeros by accessing the relevant web side. Now the challenge is whether anything can be extracted from this information and what to do with it. Its a new way of doing mathematics.

Cédric Villani: It is not entirely new. For example, the famous story of Gauss and the lemniscate was purely experimental: he computed by hand the value of the arithmetic–geometric mean of 1 and of the square root of 2, and he numerically noticed that by combining this with π, you get the length of a known curve: the lemniscate! It is a computing feat, the starting point of a new theory. Riemann's famous hypothesis, which formulates the random distribution of prime numbers in the form of a geometric localization of the zeros of the Riemann zeta function was also experimental: he computed some of the zeros, saw they were aligned on a vertical line with abscissa $1/2$, and he deduced his hypothesis... Now its true that with a computer we are on a completely different scale, we can do billions of such computations in no time... this means a new field and the necessity of a strategic reflection about it.

Pierre Cartier: I did not say that the mathematics were new, what is new are the means. Since Galileo threw a stone, what a change of scale!

Cédric Villani: It is important to be aware that interactions between computer science and mathematics, which takes place at various levels, change many things both quantitatively and qualitatively: development of new branches of logic and algorithms (polynomially computable, polynomially verifiable problems, etc.); new statistical methods, gigantic data processing techniques, artificial intelligence... All these problems stemming from computer science were well explained in a special issue of the journal *La Recherche*,[59] coordinated by computer scientist Gérard Berry.

Pierre Cartier: There are other developments we have not mentioned: proof assistants (softwares) such as Coq, for example... Briefly, proof assistants aim towards intelligent systems, not to prove theorems—that was an illusion!—but to check them.

---

[59] *Les dossiers de la Recherche* (December 2011).

Cédric Villani: Their aim is also to avoid disasters such as the crash of the Ariane rocket 5 (on June 4, 1996), because there was a division by zero in a discarded program loop... Proof assistants too change our perspective, logic, and update Hilbert.

Sylvestre Huet: The origin of Ariane 5's crash is somewhat complex... Indeed, European engineers has kept some bit of the "hard" software of Ariane 4 in one of the equipment bay cards, on grounds that there had never been problems with it. However, this new rocket was far more heavy and powerful, which enabled its flight path to be quite different at takeoff, with a shorter rectilinear part, and an earlier initial turn. Ariane 5 was doomed from the start, for as soon as it began to turn, leaving its vertical position, this bit of hardware started to go haywire, the three calculators misjudged the real altitude of the rocket, locked the motor cylinders to supposedly straighten up the trajectory when in fact they were bending it so sharply that the torsion literally cut the rocket in two. A week before its launch, I wrote an article which essentially started with: "According to the director of Aerospace, Ariane 5 has already flown thousands of times, in computer simulations; so it will be the most reliable launch ever made..." and I ended the first paragraph with "Some cheek!". Yes, they had checked the software thousands of time... except that they couldn't check it by making the rocket turn, a special manipulation would have been needed for this. As a result, they could not see that a small bit of the code was incompatible with Ariane 5's flight envelope.

Cédric Villani: There lies the very challenge of automated checking.

Pierre Cartier: The challenge does not consist in having an irrefutable proof of the four color theorem—I'd even say we don't care. But the strategy that needs to be implemented to make the proof of the four color theorem irrefutable—which can be done with Coq—requires an instruction checking system. The traffic management program of the SNCF contains 2 million lines of programming; Ariane also amounts to several million lines of programming. This order of magnitude is far higher than anything required for the hardest mathematical theorems; but on the other hand, the strategies that need to be developed are the same in all cases. In some sense, automated checking of theorems is a wonderful testing ground.

Cédric Villani: And as computer scientists say, an algorithm, a processor or a mathematical proof, they are all the same. The same logical rule underpins these three things.

Pierre Cartier: I read Thomas Hales' enthusiastic comments, who—in theory—has proved the Kepler conjecture about sphere packing. He was very happy with his proof. To give an order of magnitude, the purely mathematical part which does not use computing is 600 pages long; I attended some seminars on this topic in Princeton and at the end, the examiners gave up so the article is published with a small note: "98 % checked...". Behind this proof there is a horrid programming part—for example, there are optimizations of functions of 150,000 variables! Hales explained that these methods are going to change the way mathematics is done because proof assistant developers aim to create a mathematical reference library containing every content until at least third year undergraduate level. Starting from a nucleus containing 500 instructions in C_{++},[60] which is not much and can be controlled manually or automatically, they add all the rules of logic—Hales developed Higher Order Logic, based on Russell's[61] system. There is something here which is going to change the philosophy of mathematics because set theory no longer works in this system. On the other hand, type theory[62] does. All these challenges intermix. An old problem in category theory[63] is that the set of all sets makes no sense—everyone knows it and Russell discovered it. Despite that, at every turn, all specialists use the category of all categories. All this is logically irresponsible.

Cédric Villani: Like very large cardinals, for instance. Horrendously larger infinities than that of sets a mathematician manipulates spontaneously during his lifetime... Logically, there are no flaws, but it is questionable... even if we are forced to admit that some interesting discoveries follow indirectly from the study of these abstract concepts.

Gerhard Heinzmann: I don't know who Pierre has in mind when he mentioned specialists using the category of all categories, but I know that Yuri I. Manin conceived a category hierarchy where the totality of all categories is itself a 2-category, etc.

Pierre Cartier: To guard against such a contingency, Grothendieck formulated his version of large cardinals which he named universes, but everyone knows this is just absolute logical madness. I've thought about this because I discussed it with category theorists; in type theory, these problems are avoided. Why? I always present this ironically: what is the set of green cats? From a set theory standpoint, where a set is defined by its elements, this means that I have a big enough room where I can take the register of all green cats and check that they are all there, and that there are no nongreen cats. From a biologist's standpoint, what is a green cat? A box with a label "green cat", and each time he catches a green cat he puts it in—thats type theory. The set is not predetermined, it is evolving.

Cédric Villani: Seen from this angle, set theory is meaningless. There is a totalitarian will.

---

[60] C_{++} is one of the most widely used programming languages. It enables procedural programming, namely object-oriented programming and generic programming.

[61] Bertrand Russell (1872–1970), British mathematician and philosopher, considered to be one of the most important philosophers of the 20th century and one of the founders of modern logic. He was also a formidable polemicist, who fought for pacifism, feminism...

[62] Name given to Russell's logical system. It is close to Linnés classification system of living organisms.

[63] Category theory studies mathematical structures and their relations. Thus there is the category of groups and that of spaces, and topology partly aims to define general constructions that associate one (or several) group(s) to a space.

Pierre Cartier: Exactly, type theory is an evolving theory. The category of categories does not exist in the sense that there are no rooms where all the categories can be put at the same time, but each time I have a category I know in which box to put it. If we want to do some forecasting, I bet that in the next thirty years, there will be a radical paradigm shift! It will take time, like all paradigm shift. But this has the potential to revolutionize things. It combines both experimental mathematics and advanced computer science, and leads to logical concerns. And this is where I thing future evolutions will happen.

Cédric Villani: It is not possible to be thorough and cite every evolution but these stories clearly illustrate several concepts we have discussed: the almost cultural influence that the development of computer science is going to have on other disciplines, the practical ramifications like the problems of checking embedded softwares, interactions between theoretical and applied aspects, organizational issues…

New Paradigm, New Frontiers

Pierre Cartier: According to Hales, in thirty years time, together with our articles to journals—printed or otherwise—, we will send programs that will serve as a checking tool; the editor will then proceed with the checking and the article will be guaranteed!

Another looming change is related to the ghastly proofs that are going to accumulate. The proof of the four color theorem already sufficiently ghastly, is now under control. But for the classification of finite simple groups, specialists are not really sure it works. A proof already 10,000 pages long still needed another 800 pages, added recently, because a particular case had been omitted! And for the Kepler conjecture, there are going to be more and more pages!

Intellectually, we are facing a radically novel situation where we will be doing mathematics not to understand them, but to construct theoretical or physical models. We will build astronomical clocks that will not enable us to understand the Solar System, but will imitate its behavior and movements. We are not only facing a change in daily practice, but a complete paradigm shift. The purpose of mathematics has always been to describe and prove relations (a basic example: the three heights of a triangle passing through the same point), then to cover them up with logical links. Thus, if you approach this chain of relations from one or the other end, you see the whole of it. Traditionally, these logical links are conveyed by a plea, a line of reasoning. Henceforth, we will describe the working of a machine which will generate mechanically its own line of reasoning, but we will no longer be able to follow it step by step. And I think it is necessary to stay alert so as not to be outflanked and keep in control.

Gerhard Heinzmann: If we end up in a paradigm where we no longer understand the mathematics we do, then this would amount to blind instrumentalism. We would then be giving up an Aristotelian distinction which matters to me: the scientist differs from the experimentalist as someone wishing to understand what he uses. This new paradigm would, therefore, mean the death of the scientific paradigm.

Jean Dhombres: This outburst conjures up a new problem: should we engage in the exploration of a seemingly emerging theory on a more or less old problem, when we suspect that the work involved to establish the theory may be particularly long, and that the said theory may not have more to offer than the solution of the said problem? We could say that theoretical work itself is starting to have its own economy.

Cédric Villani: The problem of gigantic data is perhaps (I was going to say: probably, but it is best to remain prudent when predicting the future...) going to become a major focus of sciences in the future: Presently we have so many sensors and so much storage capacity that we are literally collapsing under the weight of data, data that we do not have the time to analyze, and that we do not know from what angle they should be analyzed. The emergence of statistical learning, artificial intelligence techniques, with the idea of processing a lot of data in an evolutionary way, is a sign of the times; but some would also say that it is a form of abdication in the face of complexity (leaving the machine "to understand" without trying to understand ourselves). It is important that mathematics preserves its character and its requirements, in harmony with computer science and not swallowed up by the latter.

Mathematics in the World

A Multipolar World of Mathematical Activities

Sylvestre Huet: How many mathematicians are there in the world today?

Pierre Cartier: About 100,000 in the world, of which 5,000 are in France.[64]

Sylvestre Huet: Here you are talking of mathematicians who produce new mathematics, not those who use them, are you?

Cédric Villani: But what do we mean by 'to use"? Use by applying or as an end user? There are billions of end users; if it is meant in the sense "to do calculations", all engineers, physicists, etc. have to be included.

Sylvestre Huet: In other words, more than 95 % of professional mathematicians producing new mathematics since Jesus-Christ, or since Thales—are living!

Pierre Cartier: Almost all!

---

[64] According the INSMI, there are about 3500 researchers and teacher-researchers, as well as 1500 post-docs in France (2011 figures).

Cédric Villani: And the number of researchers, all sciences taken together, is estimated to double within ten years because of the development of emerging nations like India, China, Brazil, etc.

Pierre Cartier: It is a reasonable estimate that I was able to check in situ. When I was in India a few months ago, I did not recognize the country I had seen fifteen years earlier. The same holds for South Korea, where I went in 2011. The development of these countries is spectacular. In India, there is for instance a new federal program to develop science, supported by the federal government and by the department of atomic energy (because it is the only one having any serious money). I visited several "centres of excellence"' in which the best students are selected, they are taught by the best professors, and hence a bright future is created for them! In the case of China, figures speak for themselves: in 1976, the number of official publications of the Chinese Academy of Sciences (Sinica), all disciplines taken together, amounted to about 800 pages per year. Today, mathematical publication represents tens of thousands of pages per year. Yet thirty years ago, on a scientific level, this country did not exist.

Sylvestre Huet: According to *Nature publishing*, if the curves for last ten years are extended, in 2022 China will publish as many—if not more—scientific articles as the United States, namely about 350,000 each.

Pierre Cartier: I was an Invited Professor at Austin (United States) for some years, and I saw many Chinese students. My brother, who is a sinologist, told me what a Chinese official had said to him: "We know very well that half of the students we send to the West will never return, but you know, sufficiently many return!"

Cédric Villani: And the numbers returning are increasing. Finally, this example shows that the issues at a scientific level are the same as those at a political and economic level: on the one hand, there are emerging countries who are endowing science with considerable funds and on the other, traditional developed nations— France, Russia, Germany, United States—are wondering how to keep up with this rise. What policies should be adopted? How to be efficient, given that we are in a context where we are not rolling on money, that bureaucracy is also expanding in the scientific world and the big operators—especially in France—are public, with all that this implies in terms of rules and constraints to steer our ship? The current dynamics make it very hard to stand up to competition.

Jean Dhombres: It also needs to be said that we are very badly prepared for this competition. I remember the early 1980s, at the end of the cultural revolution, when we had great difficulty making things move to enable the Chinese to come and pursue doctoral studies in mathematics in France, and also the reticences to set up mathematics classes given by the French in China, even though there is a loyalty rule in mathematics: once the student has become a master, whether in China or gone elsewhere, he remains loyal (in the good sense of the word) to the country that has contributed to his intellectual training. We mentioned earlier the influence exerted by Gottingen on Bourbakists during the interwar period: the rise of Nazism radically changed the relationship, but loyalty to Hilbert and to the spirit of Gottingen was not affected.

Sylvestre Huet: Can we present a quick overview of mathematics throughout the world: who does mathematics? Where can vibrancy be found nowadays? And on the other hand, what place does France hold in the global concert, and are we doing what is necessary to play an important role?

Cédric Villani: Since the mid-20th century, research has been dominated by the United States, in mathematics as well as in other fields. They are the ones who set the pace, in terms of production and diversification. They have significant budgets, an excellent organization and their transparency concerning public money spent in mathematics is exemplary—which is not necessarily the case in other disciplines. The American Mathematical Society is by far the most important mathematical scholarly society in the world, and it is an obvious step for all mathematicians to spend time in American universities. In my case it was Atlanta, Berkeley and Princeton. Even if you don't stay on there, its part of the education process because till now the main work has been conducted there. Three traditional European schools have managed to address the challenge of these mathematical Meccas and maintain a very important position: the French school, the German school and the Russian school before its dismantling (pre-1991).

Two significant trends are affecting current mathematical research. First, large-scale migrations of mathematicians linked to socio-political issues are being observed. In particular, an incredible decrease in mathematics standards in Eastern Europe since the collapse of the Soviet bloc.

In Russia, budgets collapsed. Hungary has completely lost the level that used to be found there. Some pockets of resistance remain, in particular in Rumania and Poland, but globally, the breakdown of mathematics in Eastern Europe is ongoing.

The other trend is the spectacular emergence—especially in terms of resources—of new powers such as China, Brazil and India, each with their own specificities. In China, the focus is on geometric analysis, on Chern's[65] influence and that of his descendants. In India, theoretical computer science is very cutting edge. In Brazil, the *Instituto Nacional de Mathematica Pura e Aplicada* (IMPA) is doing very well… Israel must also be mentioned as it occupies a prominent place in mathematics, in particular in theoretical computer science and logic.

Pierre Cartier: It must be said that Israel recuperated half of the Russian mathematicians!

---

[65] Shiing-Shen Chern (1911–2004), Chinese mathematician, naturalized American, considered to be one of the best specialists of differential topology and differential geometry of the 20th century.

Sylvestre Huet: And what about Italy, Spain, England that have traditionally been excellent mathematical centers?

Cédric Villani: In Italy, the academic fabric is slowly breaking up, though with some pockets, like Pisa where research in analysis is exceptional. As a result, a large number of excellent Italian researchers are moving to France and England, in the same way that China is "watering" the United States. In Spain, some favored field is developing very strongly, especially algebra. And we are still waiting for Africa to "wake up"... Personally, I regularly go to Dakar and other African towns as I am among those who consider that Africa will be a major academic protagonist within a few decades.

Pierre Cartier: I agree with you and I will try to go to Algeria in a few months for the same reasons.

Jean Dhombres: The difficulty, and not just any difficulty, is to avoid becoming manipulated by this or that political faction, facilitate a local development of mathematics, and not forget that the system of thesis cosupervision protects certain rights.

The Share of French Mathematics

Sylvestre Huet: This noncomprehensive catalog of mathematical research throughout the world gives the broad lines of the global mathematical scene. What about France?

Cédric Villani: The share of France in mathematical research is remarkably high. Through the quality of its research, France's share is hugely disproportionate relatively to its population as a whole, of its economic weight, and even of other sciences.

The level of French research in mathematics has been maintained due to reasons of organization, tradition, existing institutions. Some of them have been mentioned—the legacies of the Revolution, the creation of CNRS and then the 1930s when independently run, appropriate institutions were set up. These schools work well in all fields. In France, probability is as much done—and its probability school is one of the best known internationally—as analysis, algebra, geometry... In short, we do everything. At research level, we have more or less maintained our share.

Sylvestre Huet: Since the end of the Second World War, the United States holds a overwhelmingly dominant position. Europe was slowly rebuilt and its mathematics almost matches in quality that of the former. But especially, two demographical giants, China and India, are emerging extremely rapidly and it is hard to predict their evolution. So we already have a truly multipolar world in terms of mathematical activities, and France may end up with a share perhaps not proportionate to its demographic weight, but not very far. In your opinion, what can France do to continue to exist in such a world?

Pierre Cartier: Be European!

Cédric Villani: Precisely. First of all, we need European policies, an overall European vision in every sector. What is true for the economy and industry, is also true for all that is related to knowledge. And these policies should not aim to standardize, but to coordinate. Remember that Europe's motto is "Unity in diversity". We need to capitalize on German, French, Italian, traditions, on the emerging Spanish one, on the traditions of Eastern Europe… and to federate everything, with the necessary institutions, with a global vision, with coordinated policies in terms of demography, recruitment, exchanges, etc. France does not weight much against giants such as China and the United States, but Europe has the means to be on a par with them on this new world stage.

Jean Dhombres: Provided it is a genuine federation. I was scientific adviser in a French embassy, and became used to counting French representatives in the big colloquiums and specialized congresses, but also among invited academic visitors. In the sciences in general, they rarely made up more than 5 % of the total; in more or less pure mathematics, this proportion was frequently 12 %; in technical sectors, it was far less. But this was at a time when French professors spoke terrible English. In terms of research, Europe can only weigh if it integrates Great Britain. However, the latter is looking elsewhere, and British universities have a natural tendency to collaborate with their American counterparts, and find it harder doing so with French, German or Italian universities.

Pierre Cartier: I am less of a pessimist than you are.

Cédric Villani: Developing European scientific policies is going to be one of the key issues and I think that indeed we won't be able to build Europe without England. It is a nation that cannot be overlooked in financial terms and with a very successful research base, especially in mathematics. It is often taken as a model—assuredly in a questionable way, as any comparison in such subtle matters, difficult to quantify—in matters of scientific productivity. And let us not forget Switzerland. From a scientific and innovative standpoint, it is one of the most innovative countries in the world. One day it too will have to be integrated into Europe.

Jean Dhombres: The more so since despite not being in Europe, the Swiss play the game of Europe.

Cédric Villani: Hence a main focus of French mathematical development can be defined: Europe—a geographical focus. Then a temporal focus: training. This holds the key to everything. We talked earlier about the education system, and in my view, this is one of the major challenges. Governance and organizational issues are in the end easier to deal with than primary and secondary level educational problems. Yet this is where the challenges in effect lie. It is horridly complicated, but if we fail at these lower levels, the rest will be useless. If we lack raw materials, namely brains, all issues of academic organization, of governance... are nothing but empty words.

This brings us to the major issue in Europe, that of the decline of scientific vocations: no one reads science at university! Now, the issue of how to organize universities becomes secondary in a context where a mathematics course only has three or four students in attendance, perhaps even only one for which an entire laboratory has to work!

Pierre Cartier: In the Netherlands, the shortage of students in mathematics has forced universities (since 2005) to create a single unified postgraduate degree (except in the very north)—a remodeling I participated in.

Cédric Villani: Its incredible! It should be mentioned that the United States is also affected by this problem, a country that has traditionally based its strategy on importation, with a training system, of we could say "local production", which is not very successful. Presently, it is essentially importation from China since importation from Europe does not work so well. In Californian campuses (Berkeley and others), students, the majority of which are Chinese, are in dire straits economically and scientifically and are extremely dependent.

Pierre Cartier: Indeed the United States have become a very fragile country.

Cédric Villani: In terms of training it is important to progress, not just to preserve. The main challenge lies therefore at a pre-Baccalaureate[66] level.

Jean Dhombres: Should the final three high-school years be perhaps made more academic, more free, more palatable, with more choices?

Sylvestre Huet: Which foreign training models could France draw inspiration from?

Cédric Villani: Nowadays, we can compare: there are international scales enabling us to know what works, and what doesn't, in a given context. Moreover, the book coordinated by Michèle Artigue should be mentioned[67] as it suggests various directions in this matter. It is a matter where we should remain rational. Issues need to be clearly delineated, and considered over the longer term. It is the duty of politicians— together with pedagogical specialists and scientists—to investigate these issues and to do so on a European scale. I am not trying to say that we need a unified, single European system for all European countries: it is not possible, there are too many cultural, traditional specificities. Nonetheless, our approach should enable us to understand, through mutual inspiration, what in a given country is absurd and why, what works well there and why, etc. Let me take the example of repeating a year: this is considered the least efficient pedagogical practice, but it seems impossible to

---

[66]French equivalent of A levels. Translators's note.
[67]*Les défis de l'enseignement des mathématiques dans l'éducation de base*, Unesco, 2011.

contemplate doing away with it in France. We should not be afraid of expressions such as "setting goals", "result-oriented culture", and there are several points where the way of operating needs to evolve, and this evolution starts with quantifiable targets. Of course these targets are not financial ones, but concern the number of people that can be trained, the interest the can be roused among pupils, turnover percentages that need to be mastered, social mobility, etc.

Pierre Cartier: The problem can be stated by giving a few rather ruthless figures: in my mother's generation about 1 % passed the Baccalaureate; in my generation, about 6 %, in the present generation, 80 %. This is reality, ruthless and cruel. We did not respond to this increase, the problem is so huge.

Jean Dhombres: It is not for lack of trying. We should not forget what I will call "the weight of the French mentality". Mathematics are part of the cultural background of a certain elite. But ask anyone randomly about how he relates to mathematics, and the answer will invariably be: "I never understood anything in mathematics"— sometimes said with pride. If you ask the same question in Tokyo, or Cambridge, you won't get the same answer. How can mathematicians remedy this way of perceiving their subject? The old 18th century belief that mathematics prevents one from being intelligent is still there! You can find this in any writing of the time. Read Edgard Quinet, who did mathematics—he was even filled with wonder—whereas everyone around him used to say "only idiots do mathematics, because they don't see the world".

Cédric Villani: Victor Hugo used to also say that about mathematics.

Jean Dhombres: And Alfred de Musset, Alfred de Vigny… Only, they have all done mathematics during their studies, and their romantic sensitivity led them away from what they regarded as analytic thinking deprived of the heart's resources, but probably in their case this was more strongly felt because they associated mathematics to Republican thinking. It seems to me that despite romanticism having been left aside and the Republic being well established, there is still some misgivings about mathematics. It continues, despite the brilliance of the French mathematical school, despite the reputation of scientific classes in high school, despite the specialized teachers' body. I may be surprised by it, I may try to find historical continuities, I may also think that an anti-mathematica component has been part of the intellectual world since Aristotle, but I wonder whether this is not due to the basic freedom of thinking that mathematics offers (even though exchanges are essential as was correctly emphasized). Pride certainly goes with it, but that is not a reason why a mathematician would a priori wish to compete with his neighbor as soon as the latter starts talking about mathematics. He considers him his equal. Is the independence of the peaceful and universal "I" of the mathematician not contrary to the values of every type of communalism that divides us into competing trading blocs (buyers or consumers)?

Conclusion

Sylvestre Huet: It is now time to draw to a close this fascinating informal discussion on mathematics.

Jean Dhombres: We have talked a lot about freedom, and personally, I would like to emphasize that it is not limited to the freedom fairly often appropriated by mathematicians in the past in order to create. Complex numbers have been mentioned; they are still called imaginary, but we have also talked about the present world and its optimization calculus which lead to incredible precision. Greatly facilitated by computer science, in some way they enable the always necessary validation by reality. Freedom is also part of the education provided by mathematicians which at least makes it easier to master the discourse on calculations concerning tools of political governance, in the form of statistics, forecasts, risks, etc.

The topic about mathematicians' responsibility from a historical viewpoint, another strong line in our discussion, can, I think, be summarized by Galileo's approach who refused to reduce his work in mathematical physics to a hypothesis or a convenience, by associating a "reality" to it, the Earth's motion, and he was well aware that by doing so he was transforming its cosmological meaning, by the indispensable use of mathematics and of the concept of coordinates, enabling even Newton's calculations of the "fall" of the Moon. In this regard, Einstein talked of Galilean relativity.

Obviously, we are not trying to conclude that a mathematical reality exists, but have we not been able to check that mathematicians are involved in everyday reality in various ways that have been validated by history, yet distancing themselves. This is effectively possible through manipulating calculations and knowing their limits—the perfect analytic form. I had started with a description of mathematics as a dual game of abstraction and of action on reality: have we not finally been able to explain better that mathematics develop calculations, that are very real, even though they may concern objects whose reality is basically of no interest? We perhaps also recovered that which makes the great strength of Euclid's Elements, another one of our starting points, and the recognition of the meaning of the word geometry. Not only did it concern the Earth's measurement, but in particular calculations about objects only

© Springer India 2016
P. Cartier et al., *Freedom in Mathematics*,
DOI 10.1007/978-81-322-2788-5_5

some of which can be approximated from a terrestrial viewpoint. And in this sense, this geometric calculation, earlier labeled synthesis, obviously includes an element of aesthetic pleasure, and this habit of contemplation when it follows from an action, in some sense the satisfaction of a work well done. But always called into question, through wishing to extend it, reduce it, do better.

Gerhard Heinzmann: From a philosophical standpoint, our discussions can be summarized by an apparent paradox: the greater the complexity of mathematics, the further they seem to be from concrete facts. Yet, in reality, the most complex mathematics opens up the best path for describing concrete facts whose complexity is beyond the most sophisticated mathematics; it is our most efficient cultural tool to simplify our orientation in the world.

Cédric Villani: The principle of this informal discussion between people from different generations and with different views, going from one topic to the other, is I think well adapted to the multifaceted activity which mathematics is: science, art, social activity; with its own coherence and rules, but also in interaction with other human activities; with a long rich history, but always very modern.

Pierre Cartier: Our discussion has enabled us to show that mathematics makes up a rich, plentiful, living whole—at least I hope it has. We have dealt with many issues, touching on history, the meaning of mathematics, their applications, and also on the way to teach them to transmit this extraordinary legacy. Concerning many of these issues, we have gone very far and provided a unanimous answer, and the debate remains open. I hope this discussion will contribute to convince the reader that the future is not written, and that we are probably at the beginning of a fascinating and radically new era.

Appendix

Pierre Cartier's Reading List

The "orthodox" viewpoint of mathematics, according to Hilbert and Bourbaki can be found in three books:

N. Bourbaki, *Elements of the History of Mathematics*, Springer-Verlag, 1999.
J. Dieudonné, *A Panorama of Pure Mathematics, as seen by N. Bourbaki*, Academic Press, 1982.
F. Le Lionnais, *Great Currents of Mathematical Thought*, Dover, 1971.

These books dating from 1940–1960 are readable by nonspecialists (especially the third one). They give reliable (especially historical) information, but have a very strong "ideological" slant.

The development of geometry can be divided into three essential stages:

Euclid's Elements, Green Lion Press, 2002.
D. Hilbert, *The Foundations of Geometry*, The Open Court Publishing Company, 1950.
E. Artin, *Geometric Algebra*, Interscience Publishers, Inc., 1957.

As we have seen, Euclid is *the* classic. Hilbert's book, published around 1900, is a summary of all the critical work, especially of the 19th century, on Euclid's axiomatization. It is the starting point of modern axiomatization. As for Artin's book, it is the culmination of the Cartesian program of coordinate use in geometry through the application of "modern" algebraic methods of the German school of the 1920s–1930s.

For a proper perspective of mathematical challenges in the 20th century, Hilbert's Paris conference in 1900 on his 23 problems is a must. For a reprint of his lecture, see David Hilbert, "Mathematical Problems", Bulletin of the American Mathematical Society, Vol 37 No 4, pp. 407–436.

© Springer India 2016
P. Cartier et al., *Freedom in Mathematics*,
DOI 10.1007/978-81-322-2788-5

Set theory dominated mathematics throughout the 20th century. An introduction termed "naive" is given by P. Halmos, a great expositor: *Naive Set Theory*, Springer-Verlag, 1974.

Finally, two great masters, H. Poincaré and J. Hadamard, have looked at the mechanisms of mathematical inventions. The two reference volumes, recently reprinted, are:

J. Hadamard, *The Mathematician's Mind: The Psychology of Invention in the Mathematical Field*, Princeton University Press, 1996.
H. Poincaré, *L'invention mathématique*, Jacques Gabay, 2011.

Jean Dhombres' Reading List

I shall not confine myself to books in French, contrary to the French version, having chosen the topic of history of mathematics, and mathematical popularization using history as a support, and always with an original text on these mathematics. It is easy to find original texts on the Web.

On Euclid's *Elements*, see the translation in French and commentaries in four volumes by Bernard Vitrac, *Euclid's Element*, PUF, 1990–2001, and it replaces the remarkable but so old heath version.

On Chinese mathematics, for the book mentioned, see Karine Chemla, Guo Shu chun, *Les neuf chapitres, Le classique mathématique de la Chine ancienne et ses commentaires*, Paris, Dunod, 2005; and, for an incisive overall perspective, Jean-Claude Martzlof, *Histoire des mathematiques*, Paris, Masson, 1976, for which there exists a far better English version, *A History of Chinese Mathematics*, Springer, 1987. For oriental mathematics, a very interesting book is the one edited by Kim Plofker and Victor J. Kats, *The Mathematics of Egypt, Mesopotamia, Chine, India and Islam*, A Sourcebook, Princeton University Press, 2007. For Indian mathematics, a particularly interesting text is the edition of the *Tantrasangraha of Nīlakaṇṭha Somayājī* from Springer-Verlag in 2011 by K. Ramasubramanian and M.S. Sriram.

On Galileo, see Maurice Clavelin's translation, *Discours et démonstration mathématiques concernant deux sciences nouvelles*, Paris, PUF, 1995, and more specifically on the parabolic trajectory, see Jean Dhombres, "La trajectoire d'une parabole. Métamorphoses de la philosophie naturelle sous l'effet des mathématiques", XIIIe Entretiens de la Garenne Lemot, J. Pigeaud (dir), *Métamorphose(s)*, PUR, Rennes, 2010, pp. 213–241; Le jet d'eau et l'arc-en-ciel à l'âge baroque: réalisation des mathématiques, mathématisation de la philosophie naturelle et représentation. To be read is the recent issue, Galileo, by J.L. Heilbron, from Oxford University Press (2010). des phénomènes, Frédéric Cousinié, Clélia Nau (dir.) *L'artiste et le philosophe. L'histoire de l'art à l'épreuve de la philosophie au XVIIe siècle*, PUR, Rennes, 2011, pp. 151–196.

On complex numbers, see the original book, Jean-Robert Argand, *Essai sur une manière de représenter les quantités imaginaires dans les constructions géométriques*, Paris, 1806, published in 1874, see Albert Blanchard, Paris, 1971. There is an English version available: *Imaginary Quantities; Their Geometrical Interpretation*, primary source edition, 2014. See also two books by Carlos ALvarez and Jean Dhombres: *Une histoire de l'imaginaire mathématique. Vers la démonstration du théorème fondamental de l'algèbre par Laplace en 1795*, Paris, Hermann, 2011; *Une histoire de l'invention mathemématique. Les démonstrations classiques du théorème fondamental de l'algèbre jusque vers 1890*, Paris, Hermann, 2012.

For a purely mathematical presentation of Heisenberg's relations, and a reflection on mathematical writing in general, see Éric Guichard (dir.), *Écriture: sur les traces de Jack Goody*, Presses de l'ENSSIB, Lyon, 2012. To set the framework of 20th century analysis from a postgradualte level mathematics syllabus, see the historical notices in the translation of Walter Rudin's book, *Analyse réelle et complexe*, Paris, Dunod, 1998.

On the Sokal affair, the best is to come back to the book which includes Sokal's pamphlet, namely Jean Bricmont, Alan Sokal, *Intellectual Impostures*, London, Profile Books, 1998, and by the same authors, *Pseudo-sciences et postmodernisme*, Paris, Odile Jacob, 2005.

On mathematical philosophy, the analysis of Compte's thought can prove useful: (With Yann Clément-Colas and Jean Dhombres); critical edition of *Premiers cours de philosophie positive d'Auguste Comte*, Paris, Quadrige, PUF, 2007; (With Angèle Kremer-Marietti), *L'épistémologie. État des lieux et positions*, Ellipse, 2006.

On the didactics of mathematics, among several books, a recent useful publication: Alain Bronner et al., *Diffuser les mathématiques (et les autres savoirs) comme outils de connaissance et d'action*, IUFM de l'Académie de Montpellier, 2010. But also the original form given to teaching by Laplace, Lagrange and Monde, in *L'École normale de l'an III. Les leçons de mathématiques*, Paris, Dunod, 1992, now distributed by the École normale supérieure Press, together with the volumes for lessons in other disciplines.

Concerning the issue of aesthetics, see Jean Dhombres, "Qu'est-ce qu'un rendu artistique de la beauté proprement mathématique des proportions? L'analogie réalisée", in L. d'Agostino, C. Zotti, *L'espace spirituel, La pensée comme patrimoine*, Serre éditeur, 2007, pp. 51–120.

On mathematical biographies or presentations of mathematicians as a community and with direct participation of mathematicians, see the books in the series *Un savant, une époque* that I edited for Berlin (in particular on Cauchy, Hardy, Fourier, etc.), as well as the book written with my wife, *Lazare Carnot*, Fayard, Paris, 1992. Once more with my wife, we tried to restore mathematicians to their rightful place during turbulent periods, with *Naissance d'un pouvoir. Sciences et savants en France (1793–1824)*, Payot, 1989. I want to cite an original text, the first *Histoire de l'École polytechnique*, éd. critique, Paris, Berlin, 1987 and "La mise à jour des mathématiques par les professeurs royaux", in *Histoire du Collège de France*, t. 1, La création 1530–1560, A. Tuillier (dir,), Fayard, 2006, Chap. 19, pp. 377–420.

The vitality of history of mathematics in France is revealed by the catalogues of publishing houses like Vuibert or Ellipse, the books being essentially aimed at secondary school teachers; the book I edited in the framework of Irems, *Mathématiques au fil des âges*, Paris Gauthier-Villars, 1987, should be seriously updated. There are plenty of books in English.

I want to finish by mentioning two general reference works: one is N. Bourbaki's *Elements of the History of Mathematics*, Springer-Verlag, 1999: though it can be criticized for its bias, it nonetheless provides an ambitious vision; the other is *Une histoire des mathématiques. Routes et dédales*, published by Amy Dahan-Dalmedico and Jeanne Peiffer in éditions du Seuil, Paris, 1998.

Gerhard Heinzmann's Reading List

Paul Bernays, *philosophie des mathématiques* (1976), trad. Hourya Benis Sinaceur, Paris, Vrin, 2003.
This collection of Barnays' articles which contains an analysis of proof theory and of Platonism is one of the classic texts and was very influential in the 20th century. It deserves to be better known in France.

Denis Bonnay and Mikaël Cozic, *Philosphie de la logique*, Paris, Vrin, 2009.
This collection of texts and excellent commentaries by classical authors from Frege to Van Fraassen can be said to follow on Rivenc/de Rouilhan's collection.

Jean-Louis Greffe, Gerhard Heinzmann, Kuno Lorenz, *Henri Poincaré, Wissenschaft und Philosophie*, Berlin/Paris, Akademie Verlag/Blanchard, 1996.
This volume gives the most complete overall view of Poincaré's philosophy.

Caroline Jullien, *Esthétique et mathématiques. Une exploration goodmanienne*, Rennes, Presses universitaires, 2008.
One of the very rare monographs on mathematical aesthetics which contains many elements to understand this fascinating subject.

François Rivenc, Philippe de Ouilhan, *Logique et fondements des mathématiques. Anthologie (1950–1914)*, Paris, Payot, 1992.
Essential collection and commentaries on founding texts (bolzano, De Morgan, Boole, Frege, Cantor, Peirce, Dedekind, Schröder, Russell, Hilbert, Richard, König, Borel, Zermelo, Brouwer, Poincaré).

Marco Panza, Jean-Michel Salanskis, *L'Objectivité mathématique. Platonismes et structures formelles*, Paris, Masson, 1995.
A book which discusses and defends platonism without lapsing into simplistic ontologism.

Stewart Shapiro, *The Oxford Handbook of Philosophy of Mathematics and Logic*, Oxford, Oxford University Press, 2005.
Reference volume. It not only gives an overall view, but is also a useful introduction.

Cédric Villani's Reading List

Some excellent popular mathematics writings:

- Marcus du Sautoy, *The Music of the Primes: Why an Unsolved Problem in Mathematics Matters*, Harper Perennial, 2004.
- Alex Bellos, *Alex's Adventures in Numberland*, Bloomsbury, 2010.
- Simon Singh, *Fermat's Last Theorem*, Harper Perennial, 2005.
- Donal O'Shea, *Grigori Perelman face à la conjecture de Poincaré*, Dunod, 2007.

My book *Théorème vivant* (Grasset, 2012) which tries to give the reader insights into the daily life of a researcher.

A comic strip on logic: *Logicomix* (Vuibert, 2010), in which mathematicians and philosophers swing around, grappling with elusive ideals.

Henri poincaré *Science and Hypothesis*, Translated by J. Larmor, The Walter Scott Publishing Co., Ltd., 1905. Poincaré's books discuss the nature, value, and the practice of science, and are invaluable testimonies on his way of being and of working.

Scientific magazines: *Dossier pour la Science* (Jan-Mar. 2012), dedicated to mathematical problems; and *Dossier de la Recherche* (décembre 2011) which emphasizes mathematics and computer science.

My Web site: cedricvillani.org where I have gathered together some written and audiovisual documents for every type of public.

The Web Site of mathematics: http://images.maths.cnrs.fr/, with numerous suggestions, both on current issues and on longer-term topics; on a smaller scale, www.poincare.fr, the IHP's site, which is being set up.

Printed in the United States
By Bookmasters